More Brain-Powered Science

Teaching and Learning With Discrepant Events

More Brain-Powered Science
Teaching and Learning With Discrepant Events

Thomas O'Brien

National Science Teachers Association
Arlington, VA

Claire Reinburg, Director
Jennifer Horak, Managing Editor
Andrew Cooke, Senior Editor
Judy Cusick, Senior Editor
Wendy Rubin, Associate Editor
Amy America, Book Acquisitions Coordinator

ART AND DESIGN
Will Thomas Jr., Director, Cover and interior design
Cover and inside illustrations by Daniel Vasconcellos unless noted otherwise.

PRINTING AND PRODUCTION
Catherine Lorrain, Director

NATIONAL SCIENCE TEACHERS ASSOCIATION
Francis Q. Eberle, PhD, Executive Director
David Beacom, Publisher

Copyright © 2011 by the National Science Teachers Association.
All rights reserved. Printed in the United States of America.
14 13 12 11   4 3 2 1

LIBRARY OF CONGRESS CATALOGING-IN-PUBLICATION DATA
O'Brien, Thomas, 1956-
  More brain-powered science : teaching and learning with discrepant events / by Thomas O'Brien.
     p. cm.
  Includes bibliographical references and index.
  ISBN 978-1-936137-18-3
  1. Science--Study and teaching (Elementary)--Activity programs. 2. Science--Study and teaching (Secondary)--Activity programs. I. Title.
  LB1585.O29 2011
  507.1'2--dc22
                          2010047812

e-ISBN 978-1-936137-49-7

NSTA is committed to publishing material that promotes the best in inquiry-based science education. However, conditions of actual use may vary, and the safety procedures and practices described in this book are intended to serve only as a guide. Additional precautionary measures may be required. NSTA and the authors do not warrant or represent that the procedures and practices in this book meet any safety code or standard of federal, state, or local regulations. NSTA and the authors disclaim any liability for personal injury or damage to property arising out of or relating to the use of this book, including any of the recommendations, instructions, or materials contained therein.

PERMISSIONS
Book purchasers may photocopy, print, or e-mail up to five copies of an NSTA book chapter for personal use only; this does not include display or promotional use. Elementary, middle, and high school teachers may reproduce forms, sample documents, and single NSTA book chapters needed for classroom or noncommercial, professional-development use only. E-book buyers may download files to multiple personal devices but are prohibited from posting the files to third-party servers or websites, or from passing files to non-buyers. For additional permission to photocopy or use material electronically from this NSTA Press book, please contact the Copyright Clearance Center (CCC) (*www.copyright.com*; 978-750-8400). Please access *www.nsta.org/permissions* for further information about NSTA's rights and permissions policies.

About the Cover—Safety Issues: In the cartoon drawing on the cover, artist Dan Vasconcellos depicts the energy and excitement present in a school science lab. During actual school lab investigations, students should always maintain a safe distance from the teacher who is doing the demonstration. The teacher and students should wear personal protection equipment if the demonstration has any potential for bodily harm. Safety Notes throughout this book spell out when a demonstration requires that the teacher and students wear safety goggles or other protective items.

Contents

Acknowledgments vii

About the Author ix

Introduction xi

Science Education Topics xxv

Classroom Safety Practices xxix

Section 1: Welcome Back to Interactive Teaching and Experiential, Participatory Learning

- Activity 1 — Comeback Cans: Potentially Energize "You CAN Do" Science Attitudes ... 3
- Activity 2 — The Unnatural Nature and Uncommon Sense of Science: The Top 10 Crazy Ideas of Science and Challenges of Learning Science 17

Section 2: Science as a Unique Way of Knowing: Nature Of Science and Scientific Inquiry

- Activity 3 — Dual-Density Discrepancies: Ice Is Nice and Sugar Is Sweet 33
- Activity 4 — Inferences, Inquiry, and Insight: Meaningful "Miss-takes" 47
- Activity 5 — Pseudoscience in the News: Preposterous Propositions and Media Mayhem Matters 57
- Activity 6 — Scientific Reasoning: Inside, Outside, On, and Beyond the Box 71
- Activity 7 — Magic Bus of School Science: "Seeing" What Can't Be Seen 83
- Activity 8 — Reading Between the Lines of the Daily Newspaper: Molecular Magic 93
- Activity 9 — Pondering Puzzling Patterns and a Parable Poem 107

Section 3: Science for All Americans Curriculum Standards

- Activity 10 — Follow That Star: *National Science Education Standards* and True North 119
- Activity 11 — "Horsing Around": Curriculum-Instruction-Assessment Problems 131
- Activity 12 — Magical Signs of Science: "Basic Indicators" for Student Inquiry 143
- Activity 13 — Verifying Vexing Volumes: "Can Be as Easy as Pi" Mathematics 153

Contents

Activity 14   Archimedes, the Syracuse (Sicily) Scientist: Science Rules Balance and Bathtub Basics 161

Activity 15   Measurements and Molecules Matter: Less Is More and Curriculum "Survival of the Fittest" 173

Activity 16   Bottle Band Basics: A Pitch for Sound Science 189

Activity 17   Metric Measurements, Magnitudes, and Mathematics: Connections Matter in Science 201

Section 4: Science-Technology-Society (STS) and Real-World Science Instruction

Activity 18   Medical Metaphor Mixer: Modeling Infectious Diseases 223

Activity 19   Cookie Mining: A Food-for-Thought Simulation 237

Activity 20   Making Sense by Spending Dollars: An Enlightening STS Exploration of CFLs, or How Many Lightbulbs Does It Take to Change the World? 247

Section 5: Assessment to Inform Learning and Transform Teaching

Activity 21   A Terrible Test That Teaches: Curriculum-Embedded Assessment 263

Activity 22   Diagnostic Assessment: Discrepant Event or Essential Educational Experiment? 279

Appendix A
Alternative, Naive, Preinstructional, Pre-scientific, or Prior Conceptions Matter: Misconceptions, or a Rose by Any Other Name Is Still as Sweet (and/or as Thorny) 295

Appendix B
The S_2EE_2R Demonstration Analysis Form 305

Appendix C
Science Content and Process Skills 309

Research Cited 313

Index 323

Acknowledgments

As with the first book in this three-volume series, I gratefully acknowledge two categories of teachers who nurtured my curiosity and inspired me to look at nature with a sense of wonder and awe. First of all, to my former science teachers at the primary, secondary, tertiary, and graduate levels who introduced me to the FUNdaMENTALs of science in ways that "walked the talk" of research-informed best practices, I'd like to say thank you. Second, I owe a debt of gratitude to the many scientists and science educator-authors who have developed and/or refined discrepant-event-type science demonstrations. Since my decision in the late 1960s to become a teacher, I've actively analyzed numerous engaging expert science demonstrators and books on the topic (e.g., Michael Faraday's famous Christmas Lectures, Mr. Wizard's books and TV shows, Hubert Alyea's *Tested Demonstrations in Chemistry*, Bassam Shakashiri and the Institute for Chemical Education, Tik Liem's *Invitations to Inquiry,* and many more in between). I'm always amazed by how many variations on a theme are possible and how far back into the history of science our best instructional activities can be traced. With this book series, I do not claim to have created any particular demonstration. Instead, my unique contribution has been one small but creative idea—namely, to use discrepant-event activities as dual-purpose models for inquiry-based, interactive teaching-learning and as visual participatory analogies for science teacher education.

With this second book, I'd also like to acknowledge my first and most important teachers, my father (deceased, but still alive in me) and mother (who continues to be a role model for how to stay young as one grows old). We don't get to pick our parents, but I don't think I could have done better. The 1930s Depression, the accidental death of his father, and the WWII enlistment of an older brother left Dad working a full-time job as a high school junior who never finished school. Later on, to support my siblings and me with his hard-working partner, my mom, Dad often held multiple jobs at one time. Among the many lessons that my parents taught me, those that have most influenced this book series and hold special relevance to the profession of teaching are the following:

1. The best teachers (and most powerful lessons) walk the talk; they do not rely on words alone, but rather model the kinds of behaviors they want their "students" to emulate.

(continued)

Acknowledgments

2. Depending on how you choose to frame reality, work can be FUNdaMENTAL. If your "payment" for work is primarily money, you will always be underpaid. Doing good, quality work with a smile and a sense of playfulness makes people better and life more enjoyable.

3. You should always strive to do your best (given your skills and abilities at a given point in time) and then feel good about your efforts, regardless of how they rank relative to others; you should strive to beat your own former best.

4. Be serious about your efforts, but not overly serious about yourself; we're fallible—get over it. Be able to laugh and learn from your "miss-takes."

5. Put something back in the pot; to whom much is given, much is expected; and the best way to pay back a gift is to pay it forward.

With personal ownership of this book's shortcomings and grateful acknowledgment to my many "teachers" for their invaluable contributions to its quality, I invite the teacher-readers of this volume to remove the weaknesses and improve the strengths of these activities as you use them to invite your students to "stand on your shoulders" to see and travel farther.

About the Author

Dr. Thomas O'Brien's 33 years in science education began in K–12 schools, where he taught general, environmental, and physical sciences and high school chemistry. For the past 23 years, he has directed the preservice and inservice graduate-level science-teacher-education programs of the School of Education at Binghamton University (State University of New York [SUNY]). His master's-level courses include Philosophical and Theoretical Foundations of Science Teaching, Curriculum and Teaching in Science, and Elementary Science Content and Methods. He also supervises the student teaching practica. In addition, he teaches a cross-listed doctoral/post-master's educational leadership course.

Concurrent with and subsequent to earning a master's degree and doctorate in Curriculum and Instruction/Science Education at the University of Maryland-College Park, Dr. O'Brien served as a curriculum development specialist and teacher's guide editor on the first edition of the American Chemical Society's *Chemistry in the Community* (1988) textbook and as the co-author of the *New York Science, Technology & Society Education Project Teacher Guide* (1996).

As a science teacher professional development specialist, he has co-taught more than 25 summer institutes, including national programs of the Institute for Chemical Education and state and regional programs funded by grants from the National Science Foundation, the Howard Hughes Medical Institute, and the New York State Education Department, among others. He has received awards for excellence in teaching and/or service from the American Chemical Society (for National Chemistry Week programs), the New York State Association of Teacher Educators, the SUNY chancellor, and the New York State Science Education Leadership Association. These grants and awards are a reflection of collaborations with university-based colleagues and what he has learned with and from the large number of K–12 teachers he has had the privilege to serve. The *Brain-Powered Science* book series owes a debt of gratitude to these friends and funding agencies for the insights and opportunities they offered the author.

Introduction

As a science teacher-educator, I have the pleasure of addressing several audiences with my second NSTA Press book. For returning grades 5–12 preservice and inservice science teachers who have already read *Brain-Powered Science: Teaching and Learning With Discrepant Events*, welcome back. Because you are already familiar with the idea of dual-purpose activities designed for use as both inquiry-oriented *science discrepant events* and *science education analogies*, you may wish to proceed directly to the first activity, "Comeback Cans." You can also continue with this introduction to get a sense of how this book—though similar in format and approach—differs in scope from the previous book.

I'm also delighted to introduce new teacher-readers to a specially designed science education professional development book. *More Brain-Powered Science* was written so you can profit from it without having already used my previous book (although I encourage you to consider doing so). For both groups of reader-users, this introduction provides an overview and context for what this book has to offer. This book series embraces the professional development recommendations developed by NSTA (2003, 2004a, 2006, 2007a, 2007b) and ASTE (2004) and framed as a challenge in the *National Science Education Standards* (NRC 1996):

> The vision of science and how it is learned will be nearly impossible to convey to students in schools if the teachers themselves have never experienced it ... preservice programs and professional development activities for practicing teachers must model good science teaching (p. 56) ... Involve teachers in actively investigating phenomena that can be studied scientifically, interpreting results, and making sense of findings consistent with currently accepted scientific understanding (p. 59) ... Teachers also must have opportunities to engage in analysis of the individual components of pedagogical content knowledge—science, learning, and pedagogy—and make connections between them. (p. 63)

Effective science teachers catalyze and scaffold student learning by practicing both the science and art of science teaching (O'Brien 1991). Books such as *The Art of Teaching* (Highet 1950) and *The Courage to Teach* (Palmer 2007) feature the artistic, human relations, and intrapersonal aspects of teaching. Though perhaps less introspective, the craft of science education has a long history of exemplary practitioners. Scientist-teachers such as Michael Faraday (1791–1867), Thomas Henry Huxley (1825–1895),

Introduction

Richard Feynman (1918–1988), Carl Sagan (1934–1996), and Stephen Jay Gould (1941–2002) earned wide acclaim for their insightful writings, thought-provoking classes, and engaging public presentations that featured *inquiry-oriented discrepant-event demonstrations-experiments* and/or *engaging analogies* to raise fundamental questions about science. (The following sources describe a variety of pedagogically useful science analogies: Camp and Clement 1994; Gilbert and Watt Ireton 2003; Hackney and Wandersee 2002; Harrison and Coll 2008; Hoagland and Dodson 1998; Lawson 1993.)

Great scientist-teachers of any era intuitively practice science teaching as a performing art (Tauber and Sargent Mester 2007) by integrating the fun and mental aspects of intellectual playfulness. Similarly, educational philosophers from Socrates (470–399 BC) to John Dewey (1859–1952) and educational psychologists such as Lev Vygotsky (1896–1934), Jean Piaget (1896–1980), David Ausubel (1918–2008), and Jerome Bruner (1915–) have advocated for interactive instructional strategies that build on the foundation of students' prior knowledge. More recently, cognitive science research has further developed the science of science teaching as an activity that should be inquiry-based from both students' and teachers-as-learners' perspectives (APA 1997; Bransford, Brown, and Cocking 1999; Bybee 2002; Cocking, Mestre, and Brown 2000; Donovan and Bransford 2005; Mintzes, Wandersee, and Novak 1998; NRC 2000, 2007, 2010; NSTA 2004b).

This book series reflects the dual scientific and artistic nature of science teaching and is based on the following premises that bridge the gap between educational theory and practice:

1. Although cognitive learning theory and the neurobiology of emotion, perception, and cognition are in the infancy stages of development, the foundation for a science of *interactive-constructivist* science teaching has been set and is ready for use. Teaching-learning is an *interactive* process with respect to student←→teacher, student←→phenomena, and student←→student encounters. It is a *constructivist* process of learner-centered conceptual construction (rather than transmission, reception, and absorption). Given research on the science and art of science teaching, we no longer need to simply wait for great teachers to be born. Whether we are preservice student teachers, novice practitioners, competent experienced veterans, or master teachers, there are best bet steps

we can take to enhance our pedagogical content knowledge (PCK) and skills (Cochran 1997; Hagevik et al. 2010; Shulman 1986, 1987). And as we grow as professionals, we become more effective at helping our students develop as self-directed lifelong learners, our classrooms as collaborative learning communities, and our schools as exciting learning organizations.

2. Learners' often somewhat hidden prior conceptions need to be activated and challenged (Driver, Guesne, and Tiberghein 1985; Driver et al. 1994; Mintzes, Wandersee, and Novak 1998, 2000) so valid precursor ideas can be built on and extended, misconceptions can be clarified (see Appendix A), experiential and conceptual holes can be addressed, and learners can be deeply engaged in the fun work and hard play that is learning at its best. Analogically, learning can be viewed as an act of continual collaborative conceptual construction, renovation/remodeling, and expansion, where what the learner already knows (or believes to be so) influences how educational interventions are perceived and reconceived. Experientially based teaching intentionally nurtures metacognitive awareness by encouraging productive conversations between learners' prior conceptions and new experiences. *Discrepant-event activities* are powerful pedagogical tools in that their unexpected, initially anomalous outcomes stimulate the senses, catalyze conversation (both internal and external), and excite exploration that leads to conceptual conflict resolution (see *Brain-Powered Science*, pp. 350–351 for optional readings on related research).

3. Teachers play a central, catalytic role in student learning as mediated by more than their professional passions and idiosyncratic personalities (NCMST 2000; NCTAF 1996, 1997; NRC 2001a; NSB 2006; NSTA 2007a). Teachers design, implement, and revise unit-level teaching cycles and yearlong and multiyear learning progressions that address the following questions: (a) Where do we want our students to travel? [curriculum standards] (b) How can we best help them get there? [instructional strategies and sequences] (c) Where are our students relative to the target destination at any given point in a unit? [assessment system of curriculum-embedded diagnostic, formative, and summative metrics] (d) Based on this continuous cycle of "intelligence gathering," what curricular and instructional adjustments are needed to scaffold and support student learning? Curriculum-Instruction-Assessment (CIA) is an integrated system with both feedback (within a given unit)

Introduction

and feed-forward loops (that project into subsequent units); they are not separate functions performed in a consecutive linear sequence without input from changing circumstances (Enger and Yager 2001; Liu 2010; Mintzes, Wandersee, and Novak 2000; NRC 2001b, 2001c). Effective CIA focuses on "big ideas" and is somewhat holographic in that each individual learning activity, lesson, and unit contains elements of the whole.

4. "Teachers tend to teach as we were taught" is a positive, promising statement if

- we "stand on the shoulders of giants," including our own exemplary former teachers and models from the history of science;

- we become lifelong learners who engage in continual, collaborative professional development by learning with and from our science department colleagues, networking with teachers outside our school, and participating in professional associations such as NSTA that promote research-informed best practices (NSTA 2010a; Tobias and Baffert 2010); and

- we use our classrooms as action research labs and invite peers to join us in job-embedded professional learning communities and critical friends groups that collaborate on the design, implementation, and evaluation of educational experiments (Coalition of Essential Schools Northwest n.d.; DuFour and Eaker 1998; Mundry and Stiles 2009; NSTA 2010a). Effective teachers, departments, and schools reflect on their actions (Schon 1983; Appendix B in this book). Critical, collaborative analysis of our educational practices enables us to articulate and reform our often unexamined pedagogical theories in action. Rather than regression to the norm that spreads average practice, enculturation of new teachers and revitalization of experienced ones should cause progression to best practices that continuously challenge individual and institutional inertia.

With these premises in mind, both *Brain-Powered Science* and *More Brain-Powered Science* feature science *discrepant-event activities* as instructional activities that can be used for two distinct, but linked, purposes. First, the activities serve as *model inquiry-oriented science lessons* for use in preservice teacher education classes, inservice professional development settings, and the teachers' own grades 5–12 classrooms. Whether done as a hands-on exploration (HOE), interactive demonstration-experiment, or a data-based discussion, a discrepant event's surprising, often counterintuitive, outcome creates cognitive disequilibrium that causes learners to turn over or "HOE the ground" of what they already know (or what they believe to be so). Anomalous outcomes generate a need to know that motivates learners to reconsider their prior conceptions to see which ones need more "water, sunlight, and fertilizer." Equally important, discrepant events also activate and help students assess misconceptions, or "weeds that need to be uprooted," to make room for the seeds of new, more scientifically valid ideas. Purposefully puzzling activities can be used anywhere in a unit, but they are especially effective for diagnostic and formative assessment of learners' evolving mix of science conceptions and misconceptions.

Second, and unique to these two books, these same discrepant event activities serve as *visual participatory analogies for science teacher education and/or model examples*—to catalyze the teacher-as-learner's creative use of research-informed science education principles. Visual participatory analogies are a new professional development strategy in which teachers interactively participate as learners and use discrepant-event science phenomena in ways that metaphorically help them bridge the science education theory–practice gap (O'Brien 2010).

More Brain-Powered Science features some key implications and applications of cognitive learning theory and research as they relate to science education. Specifically, it focuses on interactive teaching and experiential participatory learning; human perceptions as a window to conceptual construction; learning as a psychologically active, minds-on process that depends on activating attention and catalyzing cognitive processing; and the role of prior knowledge, cognitive inertia, and misconceptions. The activities in this book will review some of these learning principles but use them as a lens to probe more deeply into other relevant curriculum-instruction-assessment (CIA) issues that are commonly explored in preservice science methods courses and inservice professional development programs.

Returning readers will note that a smaller range of science concepts (see Appendix C) is covered in this book because this book focuses more on modeling how to develop students' inquiry and process skills in the context of activities on the nature and history of science, mathematics, measurement, and science-technology-society issues and how to use assessments to inform learning and transform teaching. As such, a number of the activities in this book (e.g., #2, #5, #9, #21, and #22), though engaging discrepant events, are not classic, manipulable experiments designed to introduce specific science concepts.

Introduction

Teachers have a limited time to devote to personal learning and professional development relative to the more immediate, pressing task of preparing for what they're doing tomorrow with students. Both of my books are designed to unlock these two doors with one key. That is, teachers can increase both their science content and pedagogical content knowledge and expertise by testing out and reflecting on the impact of these activities in their own grades 5–12 classrooms as a form of job-embedded professional development (see Appendix B for a demonstration-lesson analysis form).

Organizational Structure of the Book

This book's 22 interactive, experiential learning activities (and approximately 80 related Extension activities) are clustered into 5 sections that closely parallel the NSTA Standards for Science Teacher Preparation (NSTA 2003). Professional development specialists and college-level science teacher educators may use the separate sections as a framework for a series of linked professional development sessions or in more formal, credit-bearing science methods courses. Given the range of topics covered, this book (especially when combined with its predecessor) can serve as an activity-oriented supplement to more conventional science teaching methods books (e.g., Bybee, Carlson Powell, and Trowbridge 2008; Chiappetta and Koballa 2010; Gallagher 2007; Lawson 2010). Or, if supplemented by instructor-selected readings, this book can be used in lieu of conventional methods books. Nearly every activity features both the nature of science (Abd-El-Khalick, Bell, and Lederman 1998; Aicken 1991; Bell 2008; Clough 2004; Cromer 1993; Lederman 1992, 1999; McComas 1996, 1998, 2004; NSTA 2000; Wolpert 1992) and the nature of teaching and learning (Michael and Modell 2003; Michaels, Shouse, and Schweingruber 2008; Mintzes, Wandersee, and Novak 1998). Individual grades 5–12 science teachers not affiliated with a course or professional-development program can explore the activities as nonsequential, inquiry-based lessons as linked to their instructional scope and sequence (see Appendix C). In this case, the science education themes will be encountered on a need-to-know basis in the context of regular classroom teaching.

Section 1: Welcome Back to Interactive Teaching and Experiential, Participatory Learning

This foundation-setting section is predicated on the notion that science teaching that leads to student learning is not a simple one-way process of active knowledge transmission (i.e., teaching as telling) and passive reception and absorption (i.e., learning as listening). Instead, the teaching ←→ learning dynamic system is better conceptualized as a psychologically and socially interactive, constructive process that includes both inside→ out and outside→ in exchanges between learners' internal neural networks and their external educational environments (i.e., teacher, peers, and physical and virtual phenomena). Activity #1, "Comeback Cans," uses a classic, hands-on discrepant event to invite teachers to consider how interactive teaching and participatory learning can motivate students to develop an "I can do science" attitude and come (back) to science class with anticipation and leave with regret (rather than the reverse). For teachers new to the idea of visual participatory analogies, the activity also uses a racing metaphor to raise the question of whether going faster is truly a winning strategy for maximizing student learning.

Activity #2, "The Unnatural Nature and Uncommon Sense of Science"—although not designed as a conventional discrepant event or a visual participatory analogy—helps establish the broader context and need for the rest of this book. It uses engaging visual, auditory, and physical props to help learners brainstorm a list of seemingly outrageous yet well-established core science ideas and use that list to discuss why learning science can be unnatural (or at least a somewhat unique challenge as compared to other school subjects).

Section 2: Science as a Unique Way of Knowing: Nature of Science and Scientific Inquiry

Activity #3 opens this section with a series of discrepant-event-based teacher demonstrations and hands-on explorations on the concept of density. Analogically, the argument is made that the failure of students to learn science cannot be viewed solely as a problem of students being "too dense." Instead, the failure to learn must be viewed as a system phenomenon like floating and sinking, where the educational environment must be designed to support student learning. Perhaps we load students down with too many conceptually heavy ideas to fit in the instructional time we allot. This activity and all the activities in *Brain-Powered Science* and *More Brain-Powered Science* were designed to meet the user-friendly S_2EE_2R criteria of being Safe, Simple, Economical, Enjoyable, Effective, and Relevant for both teachers and grades 5–12 students. See Appendix A in *Brain-Powered Science* for the research-based support for these criteria and Appendix B in this volume for

Introduction

a related demonstration-lesson analysis form. Five of the seven activities in this section use no-cost, inquiry-oriented paper-and-pencil puzzles (PPPs) to activate and confront common misconceptions (and conceptual holes) related to the nature of science (NOS).

The American Association for the Advancement of Science (1993), National Research Council (1996), National Science Teachers Association (2000, 2003/Standards #2 and #3), and other professional organizations agree that the NOS and inquiry should be central themes to frame science content in the context of how we know what we know (i.e., epistemology). As a field of study, the NOS is situated at the intersection of the history, philosophy, and sociology of science and cognitive psychology (McComas 1998). Every science lesson inevitably leaves students with some impressions about the NOS. Whether these impressions are scientifically valid and pedagogically motivational or not depends in part on the extent to which the teacher is intentional and explicit in focusing on the NOS as an instructional objective. An advantage of analogy-based PPPs is that they help students develop scientific habits of mind and inquiry skills without requiring prior knowledge of particular science concepts that would burden some learners and give an advantage to others. Also, given the PPPs' focus on scientific reasoning skills, these activities can be used in physical, Earth, and life science classrooms at nearly any point during the school year.

The Science Education Concepts feature of each of the seven activities in Section II also briefly reviews one of the seven principles of activating attention and catalyzing cognitive processing that were a major focus of *Brain-Powered Science*, Section III. Every activity in both books is designed in light of these principles and implicitly focuses on the NOS, but Section II explicitly features these ideas to help draw special attention to them. Effective teachers treat the NOS and learning how to learn science as a stage-setting, introductory topical unit (typically at the start of a course) and as yearlong themes that pervade all subsequent content-based units.

Section 3: Science for All Americans Curriculum Standards

Although the United States does not have an official, nationally mandated science curriculum, we have a default one based on the limited diversity of textbooks that dominate the market at any grade level or science subject area and the increasing importance and relative uniformity of high-stakes, state-mandated exams. Textbooks and tests are slowly moving in the direction called for by the *Benchmarks for Science Literacy* (AAAS 1993), the

National Science Education Standards (NRC 1996), and *A Framework for Science Education, Preliminary Public Draft* (NRC 2010). But how teachers conceive curriculum and operationally translate it into their daily instructional practices is at least as important to school science reform as new and improved textbooks and tests.

Section III includes eight activities that challenge teachers with visual participatory analogies that link curricular decisions to a compass, a horse race, chemistry "magic tricks," mathematical quandaries, historical science puzzles, measurements and molecular "magic," a homemade musical instrument, and a series of nested boxes within boxes. The professional development goal of these activities is to make the familiar look strange so that teachers can view standards-based curriculum with a fresh perspective and an eye to continuous improvement (see NSTA 2003, Standard 6). These same activities can also be used to teach grades 5–12 students science content, process skills, and habits of mind. In particular, four of the eight activities (Activities #13, #14, #15, and #17) help students develop interest and skills in measuring and mathematics that are critical to both science-technology-engineering-mathematics (STEM) careers and everyday life. Activity #14 also uses a classic historical story as an example of cross-curricular connections, and Activity #17 models how a big idea or theme (i.e., scale or powers of ten) has major implications in all fields of science and science-technology-society (STS) issues. Collectively, these activities can also provide a vehicle for teacher discussions related to 21st-century needs such as adaptability, communication skills, the ability to solve nonroutine problems, self-management, and systems thinking (Hilton 2010). References are provided for individual teachers, teacher teams, and classes who are interested in studying curriculum reform policies and practices and how they can be tested and improved on in their own classrooms.

Section 4: Science-Technology-Society (STS) and Real-World Science Instruction

Science-technology-society (STS)—including but broader in scope than environmental education—was a major programmatic thrust of NSTA in the 1970s, 1980s, and 1990s, as featured in a series of NSTA position statements (NSTA 2010b), books, and journal articles. STS is also supported by the NSTA *Standards for Science Teacher Preparation* (NSTA 2003, Standards 4 and 7), the *National Science Education Standards* (NRC 1996), the AAAS *Benchmarks* (1993), and more recent calls for science-technology-engineering-mathematics education (NAS 2007; NRC 2010; STEM Education Coalition). Debates continue about the relative efficacy of a social-issue-first, science-content-follows approach versus an approach that infuses STS concepts and real-world examples in the context of a more conventionally arranged "science first" curriculum. A one-size-fits-all approach is rarely as appropriate as having options for different students, grade levels, teachers, science fields, and schools. In the context of the 5E Teaching Cycle, STS issues can be featured during the Engage phase (i.e., to motivate interest and create a need to know) and the Elaboration or Evaluation phases (i.e., to challenge learners to become scientifically literate as they apply and extend their understanding in real-world contexts).

Activities #18–#20 provide examples of how STS can be infused into science instruction to make it more meaningful and relevant to students (see also a number of Extension activities throughout the book). The first two activities involve hands-on, minds-on, analogy-based simulations of phenomena that could not be directly experienced in a safe way (i.e., the spread of infectious diseases and the environmental effects of mining). The third activity plays off the analogical image of a lightbulb as a symbol for creativity to prompt learners to consider how informed personal consumer habits can make a big difference (i.e., think globally, act locally). Though individual and societal changes always involve trade-offs, new technologies often provide win-win options that make both economic "cents" and environmental sense. Teacher resource materials in this section help teachers incorporate more real-world STS concepts and issues into their courses whether or not they use an STS-focused textbook.

Section 5: Assessment to Inform Learning and Transform Teaching

Misconceptions about assessment that result in testing practices that fall short of best assessment practices include the following:

1. The primary purpose of assessment is to grade and rank students (and perhaps even punish laggards) with postinstruction summative tests (rather than to provide diagnostic and formative feedback to improve learning and teaching).

2. Designing fair and effective paper-and-pencil-based assessments is an easy, relatively trivial task given teachers' years of experience with taking tests (versus the difficult challenge of preparing a test where the

distribution by relative content emphasis and cognitive levels [Bloom et al. 1956] is aligned with both the intended curriculum and the actual instruction).

3. Students need to be taught content knowledge directly but not taught how to take tests (versus research that indicates that familiarity with different types of test item formats and test-taking strategies lowers student anxiety and raises their self-efficacy and actual performance).

4. Taking paper-and-pencil tests for grading purposes is, by necessity, a stress-inducing, regurgitation-type activity for students (versus an engaging challenge where students can learn new and exciting real-world applications).

5. Assessment is distinct from and planned after curriculum and instruction (versus aligned with and embedded into them as part of an integrated, intelligent CIA system in which the whole is greater than the sum of the parts).

The two activities in this section contain sample diagnostic assessments that teachers could adapt for use with grades 5–12 students. The primary purpose of these sample paper-and-pencil tests is to encourage teachers to think more deeply about the power of tests that teach students and inform their own curricular and instructional decisions and actions. They challenge teachers to consider how diagnostic, data-driven decisions to differentiate curriculum and instruction can support continuous improvement in both teachers and students. Resource books and websites on alternative assessments in science are also provided to help teachers with this crucial but very challenging aspect of teaching (NSTA 2003, Standard 8).

Appendixes

Appendix A: Alternative, Naive, Preinstructional, Pre-scientific, or Prior Conceptions Matter

Research points to the fact that in learning new science concepts, the most important variable is what the learner already knows and especially what he or she knows that isn't so. Students' alternative conceptions may be tenacious and survive conventional efforts to simply cover over them with new and improved, scientifically correct but often counterintuitive concepts. Appendix A highlights the top 10 sources of preconceptions and shows how to catalyze conceptual change toward more scientifically valid conceptions. See also *Brain-Powered Science,* Activities #20, #24, and #27–#29, and refer-

Introduction

ences in this book, such as Driver, Guesne, and Tiberghein 1985; Driver et al. 1994; Duit 2009; Fensham, Gunstone, and White 1994; Harvard-Smithsonian Center for Astrophysics (i.e., MOSART); Keeley, Eberle, and Farrin 2005; Keeley, Eberle, and Tugel 2007; Kind 2004; Meaningful Learning Research Group n.d.; Olenick 2008; Operation Physics; Osborne and Freyberg 1985; Science Hobbyist; Treagust, Duit, and Fraser 1996; and White and Gunstone 1992.

Appendix B: The S_2EE_2R Demonstration Analysis Form

Activities in the *Brain-Powered Science* series meet the criteria of being **S**afe, **S**imple, **E**conomical, **E**njoyable, **E**ffective, and **R**elevant (see also *Brain-Powered Science*, Appendix A). This checklist is designed for individual science teachers, peer coaches, lesson study groups, mentors, and supervisors to collaboratively analyze live or recorded lessons that feature discrepant-event demonstrations and experiments. The checklist provides feedback to increase both the observer's and the observed teacher's instructional effectiveness. As such, it should be used more as a catalyst for collaborative conversations than as summative assessment.

Appendix C: Science Concept and Process Skills

This appendix (in conjunction with the index) can be used to help locate activities by the featured science concept.

Activity Format

Each dual-purpose discrepant event and experiential learning professional development activity has the following standard format: Title, Expected Outcome, Science Concepts, Science Education Concepts, Materials, Points to Ponder, Procedure, Debriefing, Extensions, Internet Connections, and Answers to Embedded Questions. In *Brain-Powered Science* (pp. xviii–xxii), I discussed the purpose and rationale for each of these components, and new readers can probably deduce the purposes from the headings or by working through one activity, so only a few comments are necessary here. Given this book's focus on inquiry-based teaching-learning, the brief Expected Outcome statement and the relatively short explanations of the Science Concepts and Science Education Concepts do not need to be read before attempting a given activity. Inquiry questions embedded in the Procedure and Debriefing sections are designed to help teacher-users discover the gist of the underlying ideas by doing the activity and reflecting on the results. Probing questions are especially important when using discrepant-event activities because if preconceptions are left unexamined, such activities can lead to *new* preconceptions (even as they challenge old ones). Though these questions should also prove helpful when using the activities with grades 5–12 students, they are not intended as teacher-proof scripts. Instead, they should model and catalyze questions that the teachers-as-learners and their students will generate as they interact with the discrepant phenomena. Learner-generated questions are critical to learning as they reflect interest and cognitive engagement and provide formative feedback to both the teachers and the learners (Chin and Osborne 2008). The Answers to Embedded Questions are intentionally placed at the end of each activity to encourage teachers to approach their own professional development as an inquiry-oriented discovery, rather than a simple read-the-answers activity. Encountering new activities from the perspective of a learner who doesn't know the answers ahead of time gives teachers valuable insights into the perspectives of their own students.

Several format features are designed to serve as catalysts and resources for ongoing professional development. The Extensions are brief descriptions of related inquiry activities that are useful for independent follow-up work by teachers as a *means of assessing and extending their own knowledge of the science and science education concepts* and to support the development of units that link a series of related activities (e.g., the 5E Teaching Cycle will be a primary focus of *Even More Brain-Powered Science*). The Internet Connections provide resources for teachers (e.g., professional development links, written descriptions, and QuickTime movies of similar or related discrepant-event demonstrations and computer simulations) that, like the Extensions, are starting points for further explorations. Given the continual flux of information on the web, some links will change URLs or be dropped over time. However, most of the sites are hosted by universities, professional organizations, museums, online encyclopedias, and science supply companies that tend to have stable, long-term presences on the web. In addition to their inclusion in the text, an NSTA Press online, hyperlinked resource will allow readers to access these sites electronically and will allow for easy updating. E-learning experiences and resources are an ever-growing venue for teacher professional development and "just in time" instructional resources for teaching science across the K–16 range (NSTA 2008).

Most activities can be modeled quickly in 15–20 minutes when used as visual participatory analogies for science teacher education or as model

Introduction

science inquiry lessons with science knowledgeable teacher-learners. With instructional time so limited in most professional development settings, the activities are designed to be easy to set up, execute, and clean up. Alternatively, when used as science inquiry activities with grades 5–12 students, the activities could take up to a full class period and ideally would be placed in an integrated instructional unit of related concepts and activities.

Closing Comment

This book is based on the assumption that just as our students learn science by experiencing, thinking, writing, discussing, and doing phenomena-based science with peers, we need similar experiences to grow as teachers of science. Unfortunately, as teachers move from preservice to inservice educational settings, we often find ourselves working in insulated, isolated compartments where we neither give nor receive critical friends-type feedback and collegial support. Science progresses when the results of individual and team efforts are broadly shared, critiqued, and refined. Similarly, the science education profession progresses when we identify, confront, and are challenged to correct and learn from our preconceptions and mistakes, as well as when we share and celebrate our successes.

Rather than use this book on their own, teachers can use it more powerfully in collaborative, teachers-helping-teachers, professional development contexts (Banilower et al. 2006; DuFour and Eaker 1998; Garet et al. 2001; Loucks-Horsley et al. 1998; NRC 2001a; NSTA 2006, 2007a, 2007b; O'Brien 1992; Stannard, O'Brien, and Telesca 1994; Tobias and Baffert 2010; Yager 2005). I encourage you to use my books as vehicles to initiate or expand professional conversations and collaborations with your colleagues in your department, school district, local region, and geographically unbounded electronic networks. Although the frontline in the war against ignorance is the individual classroom, the best science teaching is not a solo enterprise, but one in which the collective, networked *we* achieves much more than the individual, isolated *me*. Career-long learning with and from our students and colleagues as we engage them in interactive, participatory, experiential learning is the hallmark of highly qualified teachers who expect and obtain the MOST from themselves (minds-on science teaching) and their students.

Science Education Topics

As with *Brain-Powered Science*, this book has two focuses: science education and science concepts. The table of contents below is organized to feature the science education themes as developed in the five sections. A second table of contents lists the science concepts alphabetically within fields of science (Appendix C). The book does not need to be used in a strict linear sequence, but rather can be explored on a need to know and use basis.

Acronyms Used in Science Education Topics

- BBS: Black Box System: A hidden mechanism is explored via observation and testable inferences.
- BIO: Biological analogies and applications are specifically highlighted.
- HOE: Hands-on Exploration: Learners working alone or in groups directly manipulate materials.
- HOS: History of Science: A story, case study, or resource from the history of science is featured.
- MIX: Mixer: Learners assemble themselves into small groups based on a specific task.
- NOS: Nature of Science: These activities focus on empirical evidence, logical argument, and skeptical review.
- PAD: Participant-Assisted Demonstration: One or more learners physically assist the teacher.
- POE: Predict-Observe-Explain: These activities use this inquiry-based instructional sequence.
- PPP: Paper-and-Pencil Puzzle: These activities use a puzzle, which is typically focused on the NOS; often a BBS.
- STS: Science-Technology-Society: The focus is on practical, real-world applications and societal issues.
- TD: Teacher Demonstration: The teacher manipulates a system and asks and invites inquiry questions.
- TOYS: Terrific Observations and Yearnings for Science: The activity uses a toy to teach science.

Science Education Topics

Section 1. Welcome Back to Interactive Teaching and Experiential, Participatory Learning

Activity	Activity Type	Science Concepts
1. Comeback Cans: Potentially Energize "You CAN Do" Science Attitudes	TD/HOE p. 3	kinetic and potential energy, friction, models, POE, BBS, NOS, and TOYS (and STS Extensions)
2. The Unnatural Nature and Uncommon Sense of Science: The Top 10 Crazy Ideas and Challenges of Learning Science	PPP (with audiovisual props) p. 17	NOS: unnatural or at least uncommon way of thinking and the "far out" nature of core concepts and theories

Section 2. Science as a Unique Way of Knowing: Nature of Science and Scientific Inquiry

Activity	Activity Type	Science Concepts and Learning Principle Modeled
3. Dual-Density Discrepancies: Ice Is Nice and Sugar is Sweet	TD TD TD HOE TD/HOE STS/BIO HOE p. 33	NOS, POE, and density (all activities), and dissolution and diffusion importance of consistent, reproducible empirical evidence, logical argument, and skeptical review to develop theories that have both explanatory and predictive value * Novelty & Changing Stimuli *
4. Inferences, Inquiry, and Insight: Meaningful Mistakes	PPP p. 47	BBS, POE, and the inferential and tentative/subject-to-revision NOS * Puzzles and Discrepant Events *
5. Pseudoscience in the News: Preposterous Propositions and Media Mayhem Matters	PPP/STS p. 57	scientific literacy and the NOS: Science and pseudoscience (astrology) are both creative, but the latter is not balanced by logical argument and skeptical review of empirical evidence. * Cognitive Connections and Meaningfulness *
6. Scientific Reasoning: Inside, Outside, On, and Beyond the Box	PPP HOE options p. 71	NOS, BBS, and POE indirect evidence, measurement skills, and inferential reasoning to discover patterns * Multisensory Experiences and Multiple Contexts *

	Activity	Activity Type	Science Concepts and "Big Ideas"
7.	Magic Bus of School Science: "Seeing" What Can't Be Seen	PPP TD option (see Activity 12) *Extensions:* BIO/STS p. 83	BBS and NOS: questions, observations, and inferences are influenced by prior understanding. Internet Connections: NOS instruments (e.g., DAST) * *Emotional Engagement, Connections, and Relevance* *
8.	Reading Between the Lines of the Daily Newspaper: Molecular Magic	HOE p. 93	NOS, POE, biopolymers, explanatory and predictive power of the atomic theory and STS/recycling * *Adequate Time for Learning* *
9.	Pondering Puzzling Patterns and a Parable Poem	PPP p. 107	NOS: pattern recognition, perceptual and conceptual biases, and science as a collective enterprise * *Psychological Rewards* *

Section 3. Science for All Americans Curriculum Standards

	Activity	Activity Type	Science Concepts and "Big Ideas"
10.	Follow That Star: *National Science Education Standards* and True North	PAD/HOE p. 119	compass directions, magnets, and NOS + STS Extension
11.	"Horsing Around": Curriculum-Instruction-Assessment Problems	PPP p. 131	problem definition and resolution and visual-spatial intelligence
12.	Magical Signs of Science: "Basic Indicators" for Student Inquiry	TD p. 143	acid-base indicators, solubility of ammonia, evaporation and NOS + Extension: STS Case Study/History
13.	Verifying Vexing Volumes: "Can Be as Easy as Pi" Mathematics	HOE/TD p. 153	volume measurement, metric units, and applied mathematics in science
14.	Archimedes, the Syracuse (Sicily) Scientist: Science Rules Balance and Bathtub Basics	TD/HOE p. 161	history and NOS, volume measurement, water displacement, and density
15.	Measurements and Molecules Matter: Less Is More and Curriculum "Survival of the Fittest"	PAD/HOE p. 173	volume measurement, significant digits, and kinetic molecular theory Extensions: STS case studies on biofuels and EcoFoam
16.	Bottle Band Basics: A Pitch for Sound Science	TD/HOE p. 189	conversion of kinetic energy to sound energy, frequency, and POE + STS Extensions
17.	Metric Measurements, Magnitudes, and Mathematics: Connections Matter in Science	PAD/HOE p. 201	metric system, powers of ten, ppm, ppb + Extensions: Biological diversity and scale effects, atomic theory, geological time, and STS/environmental issues

(continued)

Science Education Topics

(continued)
Section 4. Science-Technology-Society (STS) and Real-World Science Instruction

Activity	Activity Type	Science Concepts
18. Medical Metaphor Mixer: Modeling Infectious Diseases	HOE/MIX/STS/BIO p. 223	spread of infectious diseases by body fluids (e.g., AIDS simulation)
19. Cookie Mining: A Food-for-Thought Simulation	HOE/STS p. 237	resource conservation and waste management in mining (simulation)
20. Making Sense by Spending Dollars: An Enlightening STS Exploration of CFLs, or How Many Lightbulbs Does It Take to Change the World?	PPP/HOE/STS PAD *optional* p. 247	electrical energy conservation, compact fluorescent versus incandescent lighting, CO_2 and the greenhouse effect, and applied mathematics in science

Section 5. Assessment to Inform Learning and Transform Teaching

Activity	Activity Type	Science Concepts
21. A Terrible Test That Teaches: Curriculum-Embedded Assessment	PPP p. 263	data analysis, pattern recognition, inference making, and assessment
22. Diagnostic Assessment: Discrepant Event or Essential Educational Experiment? #1 Dueling Theories: Flat Versus Spherical Earth #2 Rooting for Plants	PPP p. 279	measurement of pre-experimental conditions and real-world examples and misconceptions related to: Evidence for a Spherical Earth History and Nature of Science Plants/BIO

Classroom Safety Practices

The discrepant-event, inquiry-based experiments in this book include teacher demonstrations, participant-assisted demonstrations, paper-and-pencil puzzles, and student hands-on explorations. In all cases, it is essential that teachers model and monitor proper safety procedures and equipment and teach students pertinent safety practices through both words and actions. Though the hands-on experiments typically use only everyday materials and household, consumer-type chemicals (e.g., water, sugar, salt, ammonia, and rubbing alcohol), teachers should consider their students' ages and particular teaching environments when deciding how to use particular activities and which safety precautions are necessary. Professional prudence, prior preparation, and practice greatly reduce the probability of accidents. Effective classroom management and safety are non-negotiable components of effective science teaching even when using very low-risk activities such as those featured in this book. Beyond these activities, consider incorporating the following best-practice safety precautions into your science teaching.

1. Always review Material Safety Data Sheets (MSDS) with students relative to safety precautions in working with hazardous materials. Chemicals purchased from science supply companies come with MSDS. These are also available from various online sites (e.g., *www.flinnsci.com/search_msds.asp*).

2. Wear protective gloves and aprons (vinyl) when working with hazardous chemicals.

3. Wear indirectly vented chemical splash goggles when working with hazardous liquids or gases. When working with solids, such as soil, metersticks, and glassware, safety glasses or goggles can be worn.

4. Do not eat or drink anything when working in a laboratory setting.

Classroom Safety Practices

5. Consider student allergies and medical conditions (e.g., latex and peanut butter allergies and asthma) when using activities that could cause a serious negative reaction.

6. Wash hands with soap and water after doing activities that involve hazardous chemicals or other materials.

7. When working with volatile liquids, heating or burning materials, or creating flammable vapors, make sure the ventilation system can accommodate the hazard. Otherwise, use a fume hood.

8. Immediately wipe up any liquid spills on the floor—they are slip-and-fall hazards.

9. Teach students that the term *chemical* is not synonymous with *toxic*, that *natural* is not synonymous with *healthy and safe*, and that chemicals they encounter on a daily basis outside of the science lab should be used in an informed manner. Scientifically literate citizens and consumers steer between the extremes of chemophobia and careless use of chemicals.

10. Science teachers should stay current on safety threats, environmental risks, and appropriate precautions as part of their ongoing, career-long professional development (CSSS Flinn Scientific; Kwan and Texley 2003; Texley, Kwan, and Summers 2004).

References

Council of State Science Supervisors (CSSS). Science Safety Guides (free downloads): *www.csss-science.org/safety.shtml*.

Flinn Scientific, Inc. Safety resources: *www.flinnsci.com/Sections/Safety/safety.asp*.

Kwan, T., and J. Texley. 2003. *Exploring safely: A guide for middle school teachers.* Arlington, VA: NSTA Press.

Texley, J., T. Kwan, and J. Summers. 2004. *Exploring safely: A guide for high school teachers.* Arlington, VA: NSTA Press.

U.S. Department of Health and Human Services: Household Products Database: *http://householdproducts.nlm.nih.gov*.

Section 1: Welcome Back to Interactive Teaching and Experiential, Participatory Learning

Comeback Cans
Potentially Energize "You CAN Do" Science Attitudes

Expected Outcome

Two seemingly identical opaque cans are set against each other in a race on a flat surface. Initially, both cans behave as expected as they roll in the direction in which they were pushed. However, the Comeback Can slows down more quickly than the other and then, most surprisingly, rather than coming to rest, it stops, reverses direction, and returns to the instructor.

Science Concepts

Learners predict-observe-explain the rolling behavior of the two cans in terms of kinetic and potential energy conversions and friction. The internal "black box" system of the discrepant can is investigated via hands-on explorations (HOEs) that include designing, constructing, and testing models that lead to a better understanding of the nature of science (NOS). When a Comeback Can is rolled in one direction, the suspended weight will cause the rubber band to "wind up," storing potential energy for subsequent release as the can comes to a stop and reverses direction. A well-designed, partially pre-wound can will, when pushed away, readily return to a point behind its initial release position on a flat, horizontal surface. If oriented properly, the can will even be able to climb up a slight incline when released without pushing.

Science Education Concepts

This homemade science toy serves as a welcome-back activity to reconnect with teachers at the start of a second class or an ongoing professional development series (or from a break during a long session). It is a both a *visual participatory analogy* and a prop to emphasize how simple *interactive experiential learning* activities CAN have students COME BACK to science class with anticipation and later leave with regret (versus coming back with regret and anticipating leaving). Teachers also consider the analogy of their work as a race in which they sometimes need to slow down and perhaps even reverse direction to maximize students' learning outcomes.

Safety Note
Take care to avoid contact with any sharp edges.

Materials

- 2 identical empty canisters such as standard 13 oz., 5.5 in. tall coffee cans.
- Each can needs 2 plastic lids and two 2 in. paper clips.
  - The can used as a Comeback Can also needs an approximately 0.25 in. wide × 3–4 in. long rubber band, a piece of thread (or twisty tie or pipe cleaner), and several 1–3 oz. weights (e.g., heavy hex nuts).

Comeback Cans

- The second can is left empty (i.e., no internal energy storage mechanism). If desired, larger (e.g., paint) or smaller (e.g., baking powder) cans or see-through, uniform diameter, empty, clean plastic peanut butter jars can be used for student models as a transparent model that reveals the "answer."

- A ball or level can be used to determine that a tabletop is flat or horizontal (a slanted or inclined plane would cause even an empty can to roll backward).

Safety Note

Some students may have peanut allergies, so make sure jars are cleaned thoroughly before using.

A simple Comeback Can may be constructed in the following sequence:

1. Drill (or poke) a hole in the center of both the metal bottom and a plastic lid of a coffee can that is large enough to push a rubber band through.

2. Attach the weights to the middle of the rubber band with a piece of thread or twisty tie.

3. Insert one end of the rubber band through the open end of the can and secure it on the outside metal end with a 2 in. paper clip.

4. Stretch the opposite end of the rubber band through the inside of the can and out through the plastic lid.

5. Secure that end with a second paper clip.

6. Place a second plastic lid on the other end of the can with the metal bottom (so the can is balanced and rolls straight). Duct tape or paper can be used to cover the two ends so the inside mechanism is not visible. For visual effect, the cans may be decorated with reflective Mylar wrapping paper if desired. See Internet Connections for alternative designs. *Note:* If the activity is done as a student team design, construct, and test project, students will need to bring in their own materials or the teacher will need to provide the materials.

Safety Note

Teacher should check these materials for safety.

Activity 1

Points to Ponder

The aim of science is to seek the simplest explanation of complex facts ... the guiding motto in the life of every natural philosopher should be, "Seek simplicity and distrust it."

—Alfred North Whitehead, English mathematician and philosopher (1861–1947)

In the matter of physics, the first lessons should contain nothing but what is experimental and interesting to see. A pretty experiment is in itself often more valuable than twenty formulae extracted from our minds.

—Albert Einstein, German American physicist (1879–1955)

image by RypeArts for iStockphoto

Procedure

When Working With Teachers

1. Welcome the teachers back from a break or previous session, and acknowledge their willingness to critically examine previously unquestioned answers about how to learn and teach science. Briefly note how the context of public accountability has many teachers feeling as if they are in a race in which their curriculum and instruction are inappropriately pulled by high-stakes, state-imposed assessments. Suggest that research-informed teaching practices internally align the separate CIA elements to be mutually reinforcing so that the metaphor of the cart (testing) before the horse (teaching) loses its negative connotation (see Activities #11 and #22). Use Einstein's quote to encourage teachers to reconsider the analogy of CIA as a race to cover content without first stimulating student interest and curiosity via engaging experimental phenomena.

2. Ask for a volunteer to assist you and give him or her a can that appears externally to be identical to yours (but lacks the internal energy storage system of your Comeback Can). Roll the cans parallel to each other (on the floor or a long table) in a race in which

the volunteer's can goes farther faster, while the instructor's can moves forward more slowly, then stops, reverses direction, and moves backward (it will even pass the point of the initial release if it was pre-wound). If time and interest permit, discuss the science via the inquiry questions listed in the When Working With Students section before proceeding to the When Working With Teachers section under Debriefing.

When Working With Students

Ask the following guided-inquiry questions. Depending on the time available, this can be done as a series of consecutive think-write (individual)-pair (dyads or triads)-share (whole-group discussion) or with a quick, whole-class poll of student ideas.

1. What do you expect to occur (predict) when you roll a can away from you, and why?

   Roll the can without the internal energy storage system to allow the learners to observe the typical outcome of a can moving in the direction of the initial push and then slowly coming to a stop. Pick up the can and put it out of the students' view (e.g., behind a box or in a sink) adjacent to a hidden, seemingly identical Comeback Can. Ask the students to explain the following:

   a. What types of forces, energy, and energy conversions are involved in the phenomenon of rolling something away from you?
   b. What variables could be changed to maximize the distance the can travels before it comes to a stop?
   c. Is it "natural" that the can (or any moving object) should decelerate and come to a stop? Are there any conditions under which a moving object would continue in a straight-line motion at constant velocity without ever stopping?

2. Emphasize that reproducibility and consistency of results are important in science. Pick up the previously hidden, partially pre-wound Comeback Can and repeat the rolling test. When the discrepant result of the can returning to you occurs, ask the following questions:

Activity 1

a. What would be the simplest explanation if you observed an object that was rolled away from you, then returned to you? How would we check if the table (or floor) was flat versus having an incline?

b. After verifying that the tabletop is level, ask the students to account for the discrepant result in this second test. When someone suggests that you used two different cans, pull out the first can and ask for a volunteer to help you run a fair race with the two cans. Ask: Assuming that we could not open these black box systems, what other tests could determine how the cans are different? (Do not give students the can to dissect or probe, but rather proceed to step #3.)

3. Hands-On Exploration: Have students work in teams to brainstorm possible, hidden, internal mechanisms that could account for the Comeback Can's unusual behavior. Challenge the teams to design, draw, construct, and test a model can that replicates this unusual behavior. Various designs will work more or less successfully. Provide a variety of relevant materials, or have students bring materials from home for a next-day activity. Students can experiment with variables such as the diameter of the can, the size and elasticity of the rubber band, the mass of the weight or how it is attached to the rubber band, and so on. After student teams test their models, discuss their observations and questions, such as the following:

a. Some models will roll alternatively forward and backward, oscillating with decreasing speed and distances covered. If energy is always conserved, why do the oscillations and movements of the Comeback Can eventually stop?

b. Does the fact that a model replicates the observable behavior of the Comeback Can provide necessary and sufficient evidence that its internal design is identical to the original system? Does a common behavior (or function) necessitate a common form?

c. Would a Comeback Can work the same way in a microgravity environment? Why?

d. How does this toy model the nature of science (NOS) and the idea of black box systems? What are some examples of black box systems you've studied in science to date? How do technological advancements allow scientists to peer into previously closed black boxes?

Debriefing

When Working With Teachers

Discuss how inquiry-oriented, interactive discrepant-event demonstrations and experiments can activate student attention and catalyze cognitive processing (i.e., the main theme of *Brain-Powered Science*). Focus specifically on how dissecting and designing, constructing, and testing of TOYS can lead to terrific observations and yearning for science. Using toys to teach science (e.g., Comeback Can) also presents science in a playful, everyday context and promotes an "I CAN do science" mindset. See the Extensions (pp. 11–12) and O'Brien (1993, 2010 [Activities #10, #11, #13, #15, #21, #22, #25, #29, and #30]); Sarquis, Williams, and Sarquis (1995); Sarquis et al. (2009, 2010); and Taylor, Poth, and Portman (1995) for discussions on how to use various science toys in teaching. Resist the teachers' requests to prematurely give them the right answer by allowing someone to dissect or open your model Comeback Can. If time does not allow them to build a model on site, challenge them to do so as homework. Current technologies always set limits on our ability to directly see the internal mechanisms of systems, from atoms to cells to organs to stars. Scientists build models based on indirect evidence to help understand complex natural black box systems. Also, the ability to do nondestructive testing (NDT) is important to a variety of real-world applications (see Internet Connections). Note: In a later session, you may wish to reveal the mechanism by sharing a see-through model (using an empty, transparent peanut butter jar as the Comeback Can).

Teachers also can be challenged to individually consider and collectively discuss (live or electronically) the following open-ended questions that relate to the analogy of teaching as a race:

- Is going faster or farther in the "wrong" direction ever desirable?
- What current schooling practices need to be stopped to better align local curriculum-instruction-assessment with national and state standards (e.g., AAAS *Benchmarks* and NRC's *National Science Education Standards*) and research?
- Is covering content faster a winning strategy for student learning?

- Are there any best practices from the past that we should consider bringing back?
- How does a particular practice integrate and create synergy between the three components of CIA?

As time permits, you also can discuss how the combination of a discrepant-event demonstration and follow-up hands-on explorations serve as curriculum-embedded assessments that can be used to activate and diagnostically assess students' prior knowledge as a launching pad for next-step learning. Prior knowledge will likely include a mix of valid experience-based, foundation-building, conceptual precursors, as well as misconceptions that need to be challenged. In preparation for developing a unit that features the Comeback Can, teachers may wish to review the research on the related science misconceptions (e.g., see Driver et al. 1994, chapters 20, "Energy"; 21, "Forces"; 22, "Horizontal Motion"; and 23, "Gravity" for an overview). As is the case with many physics misconceptions, students often adhere to more intuitive Aristotelian notions rather than the more analytical perspectives of Galileo and Newton (e.g., that the normal, expected state of moving objects is to slow down and stop rather than keep going in a straight line at a constant velocity). Misconceptions about the nature of science are discussed in Aicken 1991; Cromer 1993; Lederman 1992, 1999; Lederman and Neiss 1997; NSTA 2000; and Wolpert 1992. Teachers also can consider how they can design and analyze their curriculum-instruction-assessment as a form of action research, with ongoing feedback loops that inform the evolving CIA plans and actions (see Internet Connections: Center for Collaborative Action Research and Wikipedia).

When Working With Students

This activity can be used either in a unit on the nature of science or as an Engage-phase activity in the 5E Teaching Cycle on energy and motion that could culminate in students' testing commercial toys or designing and testing their own motion toys (see Extensions). Students also can systematically play with related computer simulations (see Java Applets on Physics and PhET in the Internet Connections). Discuss the paradox in Alfred North Whitehead's quote on page 6.

Extensions

1. *Spring-Powered Toy Cars*: Toy stores sell a variety of cars that operate in a similar fashion to the Comeback Can. The highest-quality ones (e.g., German-designed, spring-powered Darda cars and track system that are available from a variety of online sources) can be used for high-level quantitative play in high school physics classes when studying kinetic ←→ potential energy conversions and amusement park rides. (*Note:* Search the web for "amusement park physics.") These accelerating toy cars can be contrasted to battery-powered toy cars that travel at constant velocity. Alternatively, students may wish to take apart toys for young children that store and release energy in interesting ways.

2. *Smooth Sailing Science With Homemade Hovercrafts or Toy Air Pucks*: Use homemade or commercial toys (e.g., 4Kidz in the Internet Connections) to explore questions such as these: How can a system in equilibrium be either at rest or in a state of nonaccelerated, constant velocity motion? Where can one observe straight-line, constant velocity motion?

    A toy hovercraft or battery powered air puck can be pushed across a long table or clean tile floor to demonstrate Galilean motion under reduced surface friction (of course, air resistance is still a factor). Measuring tape and stopwatches can be used to make quantitative measurements for graphical analysis. Construct a small, non-latex-balloon-powered homemade hovercraft by drilling a hole in the center of a plastic soda bottle cap and gluing the cap, open end down, to an old CD. See Amateur Scientist in the Internet Connections for a large, rideable version.

3. *Can-Rolling Competitions*:

    a. Race three identical-size cans of soup—one a liquid broth, one a liquid with chunks of vegetables and meat, and one a cream-type soup—down an inclined plane or plywood ramp. Explore concepts such as the effect of mass on rate of fall, translational versus rotational kinetic energy, moment of inertia, and internal friction.

    b. Race two specially constructed cans of identical sizes and masses but with different distributions of mass; see Exploratorium SnackBook in Internet Connections.

c. Research and replicate Galileo's insights in studying gravitational effects with inclined planes (see Internet Connections).

4. *Cone Rolling Uphill as It Falls Downhill*: Two funnels glued together at their wide ends will appear to roll uphill on a two-sided ramp with an open middle section (i.e., it actually falls downhill with respect to its center of gravity). See Internet Connections: Universities of Iowa and Michigan.

5. *Rube Goldberg Energy Conversion Machines*: Video clips of complex energy transfer systems can be shown to humorously engage students' interest in the various forms of kinetic and potential energy and their conversions. Students also can be challenged to build their own systems for local and/or national competitions (see Internet Connections). Teachers can consider whether their curriculum-instruction-assessment system emphasizes fun, motivation, and interconnectedness or the frustrating, convoluted, and unnecessarily complex aspects of a Rube Goldberg system.

6. *Mystery Mailing Tubes* (see Internet Connections below) are another type of black box system that makes for a great introduction to the nature of science and scientific inquiry.

7. *Hybrid Vehicles and STS Connections*: Students can research how these cars recapture significant amounts of the energy that is normally lost as waste heat from friction during breaking. The energy is then stored in rechargeable batteries that drive supplemental electric motors, which allow hybrid cars to get more than double the miles per gallon of conventional cars.

Internet Connections

- Amateur Scientist: Vacuum Cleaner Hovercraft: *http://amasci.com/amateur/hovercft.html*. See also Leaf Blower video: *www.break.com/index/how-to-build-a-hovercraft.html*

- Center for Collaborative Action Research: *http://cadres.pepperdine.edu/ccar*

- Disney Educational Productions: Bill Nye the Science Guy: Energy, Friction, Gravity, and Motion ($29.99/26 min. DVD): *http://dep.disney.go.com*

Comeback Cans

- 4Kidz Inc.: *http://4kidzinc.com/air_puck.htm* (888-454-3799; air pucks, including a 3.5 in. diameter MicroAir Puck, item # 00910 for $9.95; powered by two AAA batteries)
- Exploratorium Snackbook: Downhill Race: *www.exploratorium.edu/snacks*
- Galileo's Inclined Plane Experiment replicated:

    http://galileo.rice.edu/lib/student_work/experiment95/inclined_plane.html

    http://illuminations.nctm.org/LessonDetail.aspx?ID=L278

    www.teachersdomain.org/resource/phy03.sci.phys.mfw.galileoplane

- How Stuff Works: Hybrid cars: *www.howstuffworks.com/hybrid-car.htm*
- HyperPhysics, Department of Physics and Astronomy, Georgia State University: Mechanics: Concept maps and explanations: *http://hyperphysics.phy-astr.gsu.edu/hbase/hframe.html*
- Java Applets on Physics: Select Motion with Constant Acceleration and Newton's Second Law: *www.walter-fendt.de/ph14e/index.html*
- Museum of Science and Industry (Chicago): Construction of a Comeback Can: *www.msichicago.org/online-science/activities/activity-detail/activities/make-a-comeback-can/browseactivities/0*
- Mystery Mailing Tubes: *http://undsci.berkeley.edu/lessons/mystery_tubes.html* and *www.bsu.edu/web/fseec/pie/Lessons2/General%20Science/Mystery%20Tube.doc*
- Nondestructive Testing (NDT) Resource Center: Commercial applications of NDT: *www.ndt-ed.org/AboutNDT/aboutndt.htm*
- PhET Interactive Simulations: Friction (rubbing of irregular molecular surfaces generates heat): *http://phet.colorado.edu/simulations/sims.php?sim=Friction*
- *Physics Demonstration: A Sourcebook for Teachers of Physics* (Julien Sprott/University of WI): *http://sprott.physics.wisc.edu/demobook/intro.htm* (see chapter 1, "Motion," 1.2 Comeback Can)
- Rube Goldberg Machines: Humor and Unnecessarily Complex Energy Conversion Systems:

    Official Website (products and competition): *www.rubegoldberg.com*

YouTube: search Rube Goldberg: *www.youtube.com*

Wikipedia: *http://en.wikipedia.org/wiki/Rube_Goldberg*

- Teacher Tube: 5th graders construct a Comeback Can: *www.teachertube.com/viewVideo.php?video_id=1471&title=Comeback_can*

- Toys in Space II: Video Resource Guide (1 MB PDF file download): *www.nasa.gov/audience/foreducators/topnav/materials/listbytype/Toys_In_Space_II.html*; Online video (38 min.): *http://quest.nasa.gov/content/rafiles/space/toys.rm*

- University of Illinois at Urbana-Champaign, Department of Physics Demonstrations: Paint Can Comeback: *http://demo.physics.uiuc.edu/LectDemo/scripts/demo_descript.idc?DemoID=790*

- University of Iowa Physics and Astronomy Lecture Demonstration: Cone Rolling Uphill: MPEG movie: *http://faraday.physics.uiowa.edu/mech/1J11.50.htm*

- University of Michigan: Physics Demonstration Catalog: Cone Rolling Uphill: *http://webapps.lsa.umich.edu/physics/demolab/Content/demo.aspx?id=318*

- Wake Forest University: Physics Demonstration Videos: Mass and Center of Mass: *www.wfu.edu/physics/demolabs/demos/avimov/bychptr/chptr2_newton.htm*

- Wikipedia: *http://en.wikipedia.org/wiki*. Search topics: action research, Galileo (e.g., his work with inclined planes), hybrid vehicles, and Isaac Newton.

Answers to Questions in Procedure, When Working With Students, steps #1–#3

1. Most students will expect a can to roll in a straight line, slow down gradually, and come to a complete stop. Although students are aware of the force that was exerted on the can by the person rolling it, most students are unlikely to think about the role of gravity and frictional forces (due to the can rubbing against the air and the tabletop or floor) in causing the can's kinetic energy

to be gradually converted into "waste" heat. Most students will know that the can will travel farther and faster if more force is exerted on the can initially, and some might suggest that the can could use gravitational potential energy if it were released down an inclined plane, but few will suggest cleaning and polishing the surface of the table to reduce friction. Most students will have the Aristotelian misconception that it is natural for moving objects to stop; they will not have the Galilean and Newtonian perspectives on inertia and frictional forces. Air-puck-type toys and video clips of experiments done in the microgravity environment of a space shuttle can be explored via the Extensions and Internet Connections activities.

2. a. An object would roll back to a person if it were rolled up an inclined plane, in which case gravity would cause it to return to a place of lower potential energy. The table's surface (or the floor) can be checked to see if it is horizontal either by using a level or seeing if an initially stationary ball will roll when placed on the surface in different locations.

   b. To account for the different behaviors of the two cans, someone will likely suggest that you pulled a trick can out from behind the box on the second run. When the two cans are run side by side in a fair test, students will surmise that something is inside the first can that makes it different. Short of actually opening up the Comeback Can to peer inside, students can compare the two cans by weighing, shaking, or drilling test holes to probe the insides of the cans.

3. Student teams may brainstorm some kind of rubber band, spring, flywheel generator-battery-motor, or other object that can convert and store kinetic into potential energy for re-release. If given sufficient time, students will discover that a variety of designs will work more or less successfully. Even if they use the same materials, there is not only one "right" design.

   a. In all cases, the Comeback Can's movement stops as energy is eventually converted to "waste" heat in the form of random molecular movements within the rubber band and the contact surfaces of the can, tabletop, or floor and surrounding

Activity 1

air molecules. Energy is always conserved, but conversion of one form to another is never 100% efficient.

b. In general, the closer the models' measurable properties and behaviors match those of the Comeback Can, the more confidence we would have that they match the internal mechanism of the Comeback Can. However, it is much easier to demonstrate that two systems are different than to prove that they are identical. Scientific models (including all theories) are typically best-fit approximations of reality that remain open to modification in light of new evidence.

c. The operation of the Comeback Can depends on gravity; without gravity, the weight would not pull down on the rubber band and allow it to become wound up. See the Internet Connections for NASA's Toys in Space program to explore how various toys operate in a microgravity environment.

d. Many of the big ideas in science can be construed as black box systems in which we rely on instruments and indirect evidence to observe and make inferential claims about constructs that we cannot observe directly with our unaided senses. Atoms, cells, the Earth's interior, and many other concepts that students have probably accepted on faith in the authority of the teacher and textbooks can be considered black box systems, especially with respect to the scientific apparatus available in schools. New science and technologies enable us to open up and look inside previously closed black boxes from atoms to the whole universe.

Activity 2

The Unnatural Nature and Uncommon Sense of Science

The Top 10 Crazy Ideas of Science and Challenges of Learning Science

Expected Outcome

Teachers (or students) are asked to generate and discuss a list of the Top 10 Crazy Ideas in Science and a related list of Top 10 Challenges of Learning Science.

Science Concepts

This brainstorming activity focuses on science as a school subject that can be mindlessly accepted or truly understood. Multiple conceptual and pedagogical discrepancies are built into this activity as teachers (or students) discuss how truly "crazy" many of the core ideas of science are when looked at from the perspective of common sense; how much of what most people "know" about science has been accepted on faith in the authority of teachers, textbooks, and tests; and the epistemological questions that humorous science cartoons, songs, and videos can help unearth about the nature of science (NOS). A theme of the *Brain-Powered Science* books is that the seemingly outrageous "big ideas in science" (e.g., the atom, cell and evolution, plate tectonics, and the conservation of matter-energy) can be made developmentally appropriate and understandable for students in grades 5–12 in light of empirical evidence, logical argument, and skeptical review. *Although the props used and conceptual focus are discrepant and elicit minds-on discussion, this activity is not a conventional demonstration-experiment. If used with middle school students, greater use of props (see the Materials list and the Internet Connections) and more time will be needed than would be necessary with senior high school students who have experienced more science courses and concepts.*

Science Education Concepts

By the end of the elementary school, students have been asked to believe a number of fantastic, counterintuitive, and perhaps seemingly crazy ideas about how the world works based on the unquestioned answers provided by teachers, textbooks, and tests (e.g., visual representations that are grossly inaccurate, such as colored, off-scale solar-system-type atoms). Though this way of knowing appears to work for *some* students if the assessments are limited to short-term recall of facts, it runs counter to the history and nature of science and to developing deep, lasting, functional conceptual understanding. Science may seem second nature to well-prepared teachers, but even teachers probably have accepted a lot of these ideas on faith. The transition from middle to high school is a good time to take stock of where we have been and where we are going and to emphasize how we know what we know (i.e., epistemology of natural philosophy or science).

....The Unnatural Nature and Uncommon Sense of Science

Effective teaching engages students in actively uncovering misconceptions, recovering valid prior knowledge, and discovering big ideas in science that predict and explain innumerable phenomena.

Materials

The Internet Connections include links to a variety of resources that can be used as discussion prompts and props:

- Posters or 2-D projected images and/or 3-D physical models of some of the big ideas, such as the solar system (off-scale), the cell and DNA double helix, and the atom and the periodic table
- Science cartoons that humorously highlight the content and nature of science
- Science songs or videos that creatively depict science concepts

Procedure

1. Project or distribute handouts that contain one or more of the quotes on pages 20–21 or science cartoons, or play video clips or audio tracks of science songs. The latter can be used as humorous discussion prompts to catalyze thinking about the nature of scientific processes, concepts, and theories. Projected images, posters, and physical props (models) of representative big ideas also may be displayed or distributed to help the brainstorming in step #2b and #2c.

2. Ask the learners to get into groups of three or four to address the following questions and tasks:

    a. Do you agree with the Points to Ponder quotes? Why or why not?

    b. Brainstorm a list of concepts and theories from astronomy, biology, chemistry, Earth science/geology, and/or physics that seem to be a lot to swallow or that you have accepted on faith in the unquestioned authority of teachers, textbooks, and tests.

    c. Based on your interaction with the representative images and models, do you think that the core concepts and theories of science are just common sense? If you had to defend

Points to Ponder

"There's no use in trying," she said. "One can't believe impossible things!" "I dare say you haven't had much practice!" said the Queen. "When I was your age, I always did it for half-an-hour a day. Why, sometimes I've believed as many as six impossible things before breakfast."

—Lewis Carroll (1832–1898), mathematician and author, *Alice's Adventure in Wonderland* (1865)

Science is arguably the defining feature of our age; it characterizes Western civilization. Science has never been more successful nor its impact on our lives greater, yet the ideas of science are alien to most people's thoughts ... many people accept the ideas of science because they have been told that these ideas are true rather than because they understand them. ...

Non-scientists do not in general have ... confidence, nor do they have familiarity with scientific thinking. For example, only about 5 per cent of Americans have been found to be reasonably scientifically literate, even though about half the bills before Congress involve either science or technology ... Science is one of humankind's greatest and most beautiful achievements and for its continuation, free and critical discussion ... is ... essential.

—Lewis Wolpert, *The Unnatural Nature of Science* (1992)

The Unnatural Nature and Uncommon Sense of Science

> **Points to Ponder** (continued)
>
> *Science is different from many other human enterprises ... in its passion for framing testable hypotheses, in its search for definitive experiments that confirm or deny ideas, in the vigor of its substantive debate, and in its willingness to abandon ideas that have been found wanting [p. 263] ... But the tools of skepticism are generally unavailable to the citizens of our society. They're hardly ever mentioned in the schools, even in the presentation of science, its most ardent practitioner [p. 77] ... at the heart of science is an essential balance between two seemingly contradictory attitudes—an openness to new ideas, no matter how bizarre or counterintuitive, and the most ruthlessly skeptical scrunity of all ideas, old and new. This is how deep truths are winnowed from nonsense. The collective enterprise of creative thinking and skeptical thinking, working together, keeps the field on track [p.304] ... Both skepticism and wonder are skills that need honing and practice. Their harmonious marriage within the mind of every schoolchild ought to be the principal goal of public education. [p. 306]*
>
> —Carl Sagan, *The Demon-Haunted World: Science as a Candle in the Darkness* (1996)

these core scientific concepts and theories in a court of law—where the three criteria of *empirical evidence, logical argument,* and *skeptical review* were required to win—could you make a solid case for science as being reasonable and defensible? Or do you think that the jury would find them to be impossible ideas (or at least highly improbable)?

3. Move the discussion from small groups to the whole class, and record students' collective ideas on a blackboard or whiteboard. Additional ideas can be pulled from the Top 10 Crazy Ideas in Science list located in the Answers to Embedded Questions on page 27. Discuss how sometimes what passes for understanding

in school science are "just-so stories" (like Rudyard Kipling's *Just So Stories*) that we mindlessly believe rather than understand. This need not and should not be the case!

4. If time permits, brainstorm a list of characteristics of science that make it challenging to learn and truly understand. See the Top 10 Challenges of Learning Science list in the Answers to Embedded Questions on page 28. Discuss how learning science for understanding involves some unique challenges, even for very intelligent and motivated students.

Debriefing

When Working With Teachers

Challenge teachers to remember what it felt like to be a novice student in science, what science looks and feels like to an outsider, and to what extent they themselves truly understand what they think they know about key science concepts and theories and their underlying epistemology. The history of science contains numerous cases of the ongoing evolution of ideas and the survival of the fittest ideas, including cases where entrenched, commonsense conceptions are eventually shown to be misconceptions. Science cartoons and jokes often are so engaging because they are discrepant in that they challenge our assumptions about both what and how we know and get us to look at things from different perspectives. For this reason and because of their motivational effects, humorous approaches can be effective in promoting understanding of FUNdaMENTAL science (see Internet Connections).

When Working With Students

Middle and high school students will not generate a list as lengthy as the one in the Answers to Questions in Procedure. It is enough if the class generates a few key ideas in several disciplines. Simply discuss the ideas offered with the whole class and challenge them to keep you honest with respect to the criteria of *empirical evidence, logical argument,* and *skeptical review* as big ideas are introduced throughout the science course. Understanding (versus believing) science implies being able to follow the line of argumentation and implications

The Unnatural Nature and Uncommon Sense of Science

(e.g., predictions and applications) of its core ideas. When meaningful learning is supplanted by blind, subservient dependency on and addiction to the authority of teachers, textbooks, and tests, the nature of science is misrepresented and the educational needs of a democracy are subverted. Being upfront with students about some of the unique challenges of learning to do science and directly teaching them how to learn science provides both emotional support and cognitive scaffolding for students' entry into the somewhat foreign culture of science.

Extensions

1. *The Uncommon Sense and Unnatural Nature of Science*: Teachers may study individually or form discussion groups based on one or more of the following books. The nature, history, and philosophy of big ideas in science (as distinct from the pseudoscience featured in Activity #5), as well as biases and limitations of human perception and cognition (see *Brain-Powered Science*, Section 2), make excellent themes for ongoing teacher professional development. The Internet Connections contain additional resources.

   - Agin, D. 2006. *Junk science: How politicians, corporations, and other hucksters betray us.* New York: Thomas Dunne Books.
   - Atkins, P. 2004. *Galileo's finger: The ten great ideas of science.* New York: Oxford University Press.
   - Cromer, A. 1993. *Uncommon sense: The heretical nature of science.* New York: Oxford University Press. See also: Cromer, A. 1997. *Connected knowledge: Science, philosophy, and education.* New York: Oxford University Press.
   - Ehrlich, R. 2004. *Eight preposterous propositions: From the genetics of homosexuality to the benefits of global warming.* Princeton, NJ: Princeton University Press. See also: Ehrlich, R. 2002. *Nine crazy ideas in science.* Princeton, NJ: Princeton University Press.
   - Gilovich, Thomas. 1993. *How we know what isn't so: The fallibility of human reason in everyday life.* New York: Free Press.
   - Haack, S. 2003. *Defending science—within reason: Between scientism and cynicism.* Amherst, NY: Prometheus Books.
   - Hellman, H. 1998. *Great feuds in science: Ten of the liveliest disputes ever.* New York: John Wiley & Sons.

Activity 2

- Kida, T. E. 2006. *Don't believe everything you think: The six basic mistakes we make in thinking.* Amherst, NY: Prometheus Books.
- Macknik, S. L., and S. Martinez-Conde. 2010. *Sleights of mind: What the neuroscience of magic reveals about our everyday deceptions.* New York: Henry Holt. *www.sleightsofmind.com* (includes multimedia clips).
- Morowitz, H. 2002. *The emergence of everything: How the world became complex.* New York: Oxford University Press.
- Sagan, C. 1996. *The demon-haunted world: Science as a candle in the dark.* New York: Ballantine Books.
- Shermer, M. 1997. *Why people believe weird things: Pseudoscience, superstition, and other confusions of our time.* New York: W.H. Freeman.
- Wolpert, L. 1992. *The unnatural nature of science: Why science does not make (common) sense.* Cambridge, MA: Harvard University Press.
- Wolpert, L. 2006. *Six impossible things before breakfast: The evolutionary origin of belief.* New York: W.W. Norton.
- Wynn, C. M., and A. W. Wiggins (with cartoons by S. Harris). 1997. *The five biggest ideas in science.* New York: Barnes & Noble Books.
- Wynn, C. M., and A. W. Wiggins (with cartoons by S. Harris). 2001. *Quantum leaps in the wrong direction: Where real science ends ... and pseudoscience begins.* Washington, DC: Joseph Henry Press.
- Youngson, R. 1998. *Scientific blunders: A brief history of how wrong scientists can sometimes be.* New York: Carroll & Graf.

2. *Creative Comic Strips Sell Science to Students*: Challenge individual students or teams to use the MakeBeliefsComix website (see Internet Connections) to make their own science-based comic strips. This site allows students to choose from an array of characters and facial expressions, then select the text that the characters speak. If desired, students also can be asked to create a consensus rubric for evaluating the strips based on both the science content and the humor. Of course, students with drawing talent do not need the support of a site like this one. Other alternatives include writing short essays on the science behind published science cartoons and using the drawings from everyday comic strips and inserting alternative words to convey a science concept or STS/environmental message.

The Unnatural Nature and Uncommon Sense of Science

Internet Connections

- AllPosters.com (search: Science): *www.allposters.com*
- Centre for Science Stories (HOS): *http://science-stories.org*
- History of Science Society: *www.hssonline.org/main_pg.html*
- International Society for the History and Philosophy of Science: *www.hopos.org*
- International History, Philosophy and Science Teaching Group: *http://www1.umn.edu/ships/hpst*
- *Internet History of Science Sourcebook:* Contains numerous links from ancient to modern times: *www.fordham.edu/halsall/science/sciencesbook.html*
- Jokes and Science: *www.juliantrubin.com/sciencejokes.html*
- MacTutor History of Mathematics: *www-history.mcs.st-and.ac.uk*
- MakeBeliefsComix! Make your own comic strip generator: *www.makebeliefscomix.com*
- PBS Teachers Resource Roundups: Developing Scientific Thinking (PDF download): *www.pbs.org/teachers/resourceroundups*
- Pseudoscience Sites: See Activity #5 for websites
- Sciencemall-usa (posters): *www.sciencemall-usa.com*
- Science Posters Plus: *www.super-science-fair-projects.com/science-posters-plus.html*
- Sources of Science Cartoons (check for copyright and permission to use):

    Mark Anderson: *www.andertoons.com/cartoons/science*

    Ashleigh Brilliant (*Brilliant Thoughts, Pot-Shots*, and more): *www.ashleighbrilliant.com*

    Nick Downes: *http://nickdownes.com* (see also the books: *Big Science* and *Whatever Happened To 'Eureka'?*)

    Benita Epstein: *www.benitaepstein.com*

    CartoonStock/science: *www.cartoonstock.com/directory/s/science.asp*

    Cartoonist Group: Science: *www.cartoonistgroup.com/bysubject/subjectcartoonists.php?sid=1020*

Activity 2

John Chase: Chasetoons Science Humor: *www.chasetoons.com/schum.html*

Randy Glasbergen, *The Better Half* (and more): *www.glasbergen.com*

Sidney Harris at Science Cartoons Plus: *www.sciencecartoonsplus.com* (published multiple compilation cartoon books on science, education, medicine, psychology, etc.)

Nick D. Kim *Nearing Zero*: *www.lab-initio.com* and *www.nearingzero.net/index.html*

Gary Larson, *The Far Side*: *www.thefarside.com*

John McPherson, *Close to Home:* Physics: *www.physlink.com/Fun/McPherson.cfm* and *www.gocomics.com/closetohome*

Mark Parisi, *Off the Mark*: *www.offthemark.com/science/science.htm*

Physlink.com: Physics and Astronomy Fun: *www.physlink.com/FUN/index.cfm*

Tom Swanson: *http://home.netcom.com/~swansont/index.html*

Bob Thaves: Frank & Earnest: *www.frankandernest.com*

- Sources of Science Songs:

    Dr. Chordate (various artists): *www.tranquility.net/~scimusic/notochordsproducts.html*

    Greg Crowther's Science Song Music: *http://faculty.washington.edu/crowther/Misc/Songs*

    Math and Science Song Information, Viewable Everywhere (MASSIVE) database: *www.science-groove.org/MASSIVE*

    Mike Offutt Science CDs (for sale): *www.teachersource.com/Chemistry/ChemistryResources/Songs%20By%20Mike%20Offutt.aspx*

    Neuroscience For Kids: Select: Experiment: Brain Songs: *http://faculty.washington.edu/chudler/neurok.html*

    Science Song Links: *www.haverford.edu/physics/songs/links.html*

    Songs for Teaching: Science Songs: *www.songsforteaching.com/sciencesongs.htm*

    WIRED Science: Top 10 Scientific Music Videos: *www.wired.com/wiredscience/2009/07/sciencemusic*

- Science Humor—It's Alive (cartoons, video clips, and more): *www.sciencehumor.org*

The Unnatural Nature and Uncommon Sense of Science

Answers to Questions in Procedure, step #2b

Top 10 Crazy Ideas in Science (Examples)

- *Astronomy*: big bang theory and the scale of astronomical time and distance (powers of ten)
- *Biology*: emergent properties and complexity (abiogenesis: nonliving atoms and molecules → living organisms → biosphere); the cell theory and the genetic code (i.e., Each somatic cell has all the information to construct a clone, and all the trillions of cells in a human originate from one fertilized egg cell, except microbes, which actually outnumber the human cells and exist as unique ecosystems within our bodies!); evolution; and the interrelatedness of all life
- *Chemistry*: conservation of matter and infinitesimally small, innumerable, indestructible, mobile atoms (models from billiard ball → solar system, with most of the volume being truly empty and nearly all the mass concentrated in a tiny nucleus → wave and particle duality and quantum mechanics) and molecules (mole = 6×10^{23}); the periodic table and emergent properties (~100 types of atoms → millions of types of molecules)
- *Earth Science/Geology*: geological time (Earth = ~4.6×10^9 years old); different layers of the Earth; plate tectonics and the continuous, depth-varied nature of the atmosphere; hydrosphere and lithosphere and their global interactions as the system that sustains the biosphere
- *Physics*: inertia and the laws of motion and universal gravitation; conservation of energy; invisible electromagnetic radiation (and its interaction with matter); speed of light, quantum mechanics; relativity, $E = mc^2$; and the fundamental unity (cosmos) within diversity (chaos)

Top 10 Challenges of Learning Science
(adapted from Comer 1993; Wolpert 1992)
Scientific processes, concepts, and theories ...

1. often are counterintuitive and cannot always be acquired by simple, haphazard, non-instrument-assisted observations of nature. Science challenges us to continually reconsider what we know that may not be (or isn't) so. In most cases, science relies on intentionally designed, controlled, technology-aided experiments that manipulate or otherwise test nature in controlled settings, in modified time frames, and with data triangulation from multiple trained observers following similar best practices. However, in the case of some research in certain scientific disciplines (e.g., archaeology, evolutionary biology, paleontology, astrophysics, and cosmology), "nature has already run the experiments." In these cases, scientists collect and weigh the evidence and propose and skeptically review logical arguments that result in explanatory theories that predict new discoveries and/or make better sense of previous data.

2. often are outside of everyday experience because of the scales of size, distance, time, or speed involved in seeing or otherwise directly sensing the phenomena (relative to human sensory limitations and the technological means of extending these limitations; see Activity #17).

3. often explain the familiar in strange terms; some phenomena may not be translatable into everyday language, metaphors, and concrete models (e.g., quantum theory).

4. require rigorous objective and logical proof based on interconnected and paradigm-consistent analyses and argumentation (including complex mathematics, statistics, and probability theory) that are linked to *experimentally falsifiable* theories.

5. require iterative cycles of creative/speculative and critical or skeptical thinking with successive, evolutionary, asymptotic approximation toward truth (versus Truth).

6. are collaborative, connected, and cumulative across individuals, disciplines, nations, and time. Students cannot be expected to leap tall buildings of hundreds or thousands of years of

The Unnatural Nature and Uncommon Sense of Science

interdisciplinary, international, and intergenerational scientific thought in a single bound in a few pages of a textbook or a short unit of instruction.

7. are judged by their explanatory power and scope; simplicity, economy, and parsimony; fruitfulness in raising and answering new questions; and testability and empirical support.

8. were not systematically developed for most of 100–200,000 years of human evolution or even the past 6,000–10,000 years of our recorded history. The relative newness of science stands in contrast to problem-solving technology that played a highly adaptive function in the success of our species from the first members of the *Homo* genus that made tools and passed on the craft of how to use and construct them to their offspring.

9. deal exclusively with the natural world of matter and energy versus the possible supernatural world of souls, spirit, and God. Science is agnostic and naturalistic in its methodological practices and explanatory and predictive theories. It cannot address teleological issues of ultimate purposes or design. Also, though science can study the evolutionary origins of ethics and morality and provide necessary information that is relevant to individual and societal choices and decisions in STS-related matters, it cannot claim the authority of knowing what is morally right or best when it comes to matters of values.

10. requires repeated, direct, guided encounters with phenomena to fully appreciate the phenomena; science isn't a spectator sport. Though we stand on the shoulders of giants, we cannot understand the nature and key principles of science by merely reading about scientific discoveries of others that came before us. Science is an action verb!

Section 2:
Science as a Unique Way of Knowing: Nature of Science and Scientific Inquiry

Dual-Density Discrepancies
Ice Is Nice and Sugar Is Sweet

Expected Outcome

Seemingly identical crystalline cubes (ice versus halite crystals) are observed to either sink or float in two seemingly identical clear, colorless liquids (ethanol and water). Also, a handful of sugar cubes are observed to sink in a container of hot water; most dissolve, but several unexpectedly float to the top.

Science Concepts

Density and buoyancy: An insoluble solid will sink in a liquid if its density is greater than the density of the liquid (i.e., unless the solid's mass distribution over a relatively large surface area allows it to be supported on the skin of a liquid with a high surface tension). Floating and sinking are system-level phenomena that depend on how two substances interact. The activities also address the nature of science, including the importance of consistent, reproducible empirical results (or evidence), logical argument, and skeptical review to develop theories that have both explanatory and predictive value. Science is more than a body of knowledge and a narrowly conceived, cookbook-like single method; it is a way of constructing knowledge. Consideration of *how* we know what we know is as important as *what* we know.

Science Education Concepts

Science discrepant events, magic tricks, and a little visual subterfuge can be used to challenge students' abilities to make careful empirical observations, logical arguments, and skeptical review (i.e., predict-observe-explain) while simultaneously probing their conceptual understanding of FUNdaMENTAL concepts such as density. Even simple experiments have numerous variations that teachers can explore to deepen their own understanding as preparation for engaging their students in scientific inquiry that seeks to explain apparent discrepancies. This discrepant-event activity also demonstrates the pedagogical principle of the power of *novelty and changing stimuli* in *activating attention and catalyzing cognitive processing* in learners. The brain's ability to selectively notice changes in our environment (and, conversely, to deselect or tune out unimportant stimuli that do not change) is linked to both the survival of our species over evolutionary time and our lifelong ability and desire to seek out and learn from new experiences. Activities #13 and #14 in *Brain-Powered Science* emphasize this same pedagogical principle. Optimally, schools are places where students experience new dimensions of sights and sounds and ever-expanding horizons of ideas!

For a *visual participatory analogy*, you can consider whether the problems of retention and transfer of learning in science education

Dual-Density Discrepancies

are that some students just will not understand, the cognitive load or density of concepts (i.e., the ratio of the mass of new concepts to instructional space or time) is too great for students at a given grade (or developmental) level to make sense of the concepts, or the curricular and instructional scaffolding are insufficient to support the weight of the assessments.

Materials

Demonstration Experiment #1
- Halite (sodium chloride crystals) used for cleavage experiments in Earth science classes (available from many science suppliers), cut to ice-cube-size pieces
- Ice cubes
- Water
- Ethanol (or any alcohol or alcohol solution with a density less than 0.92 g/ml); drugstore-variety antiseptic alcohols (70–90% ethyl or isopropyl alcohol) will work.
- Ice chest or cooler
- Clear, colorless plastic cups
- *Optional:* wax or paraffin

Demonstration Experiment #2
- Sugar cubes
- NewSkin liquid bandage
- Hot water
- Tall cylinder (e.g., empty 3- or 4-ball plastic tennis container)
- *Optional:* If desired, fake Styrofoam "sugar" cubes can be used as an extra discrepancy, and music that features the word *sugar* can add some levity (e.g., "Just a Spoonful of Sugar" from the film *Mary Poppins*).

Safety Note
Alcohol solutions are flammable and should be kept away from fire, heat, sparks, and electricity.

Activity 3

Points to Ponder

Men learn while they teach.
—Lucius Annaeus Seneca (the Younger), Spanish-born Roman statesman and philosopher (4 BC–65 AD)

To teach is to learn twice.
—Joseph Houbert, French essayist (1754–1824)

The mediocre teacher tells. The good teacher explains. The superior teacher demonstrates. The great teacher inspires.
—William Arthur Ward, American author (1921–1994)

Procedure

Demonstration Experiment #1: Ice Is Nice, But Think Twice. It's Not My Fault, It's Salt!

1. *Before* the learners arrive in the classroom, the instructor should fill a clear, colorless plastic cup three-quarters of the way to the top with ethanol and place it out of sight in a sink.

2. Introduce the activity by asking the learners to predict what will occur when an ice cube is dropped into water. Elicit their ideas (most will predict it will float) and ask for their reasons. *Note: At a later time* in the unit, point out that water actually displays an unusual liquid-solid phase interaction. That is, the solid phase of water is actually less dense than the liquid form. This is in contrast to the behavior of most liquids, which become denser as they solidify (i.e., molecules lose heat energy to the environment, slow down, and move closer together). Demonstrate this more "normal" behavior with solid butter or margarine, which sink in their liquid.

3. Ask the learners to observe as you take an empty plastic cup, place it out of sight in the sink, and turn the faucet on to fill it.

Dual-Density Discrepancies

4. Pick up the plastic cup that previously had been filled with ethanol (most students will not notice your subterfuge in switching cups) and place it on the desktop. Pull out a standard ice cube from an ice cooler and drop it into the ethanol.

5. As the ice cube sinks in the ethanol, feign surprise at the discrepancy and ask students to explain the discrepancy. If necessary, lead the learners to question the assumption that the clear, colorless liquid is necessarily water (e.g., Did they really see you pour water into the cup?). Remove the ice cube from the ethanol so that the undiluted ethanol can be reused later.

6. After you elicit a variety of ideas, repeat the experiment, but hold the cup up as you visibly fill it with tap water. The same ice cube used before will float in the tap water as expected.

7. Indicate that repeatability of experimental results is important in science, so perform the experiment one last time. This time, pull a halite crystal that looks just like an ice cube (cut to approximately the same size as a standard ice cube) out of the ice chest. Pour a second cup of water and drop the halite into the water. As the halite cube sinks in the water, elicit explanations for this additional discrepancy. Again challenge the idea that students saw you use an ice cube.

   Note: Do not leave the halite crystal in the water too long, as it will readily dissolve in water. Ask: Can a solid be said to be a floater or sinker independent of the liquid in which it is placed? Why or why not? If desired, additional examples include the following:

   a. Paraffin wax (d = 0.85 g/ml) in water versus in ethanol (and methanol or isopropanol; all three of these alcohols have a density of about 0.79 g/ml).
   b. The behavior of an ice cube (d = 0.92 g/ml) in corn oil (d = 0.91 g/ml) versus corn syrup (d = 1.38 g/ml)

Demonstration Experiment #2: Sugar Is Sweet, But When It Floats, It's Neat

1. An hour or more *before* class, pretreat several sugar cubes by coating them with New-Skin liquid bandage. Sugar-cube-size pieces of Styrofoam are another kind of fake sugar cube.

Activity 3

2. Ask the learners to predict how sugar cubes will behave in water. Place a handful of sugar cubes (including several of the pretreated cubes) in a tall cylinder of warm or hot water. If possible, play the song "Just a Spoonful of Sugar" from the film *Mary Poppins* for a little levity. Ask: Are the changes we observe when we place sugar in water considered physical or chemical changes? Why? What are the classic signs of chemical reactions? Can you think of examples in which solids both dissolve and chemically react when placed in liquids?

3. The learners will observe that the sugar cubes will drop to the bottom and begin to dissolve as predicted. *Note:* If fake sugar cubes (Styrofoam) are used, they will instantly float and can be pulled out and passed around as obvious "trick" sugar cubes. Elicit ideas as to how one could speed up the dissolution of sugar, and discuss why such ideas work at a molecular level. Within a few minutes of the start of this discussion, and after the untreated cubes have already dissolved, several of the pretreated cubes will mysteriously rise to the top and float.

4. Elicit possible explanations as to what might be going on. Distribute several of the pretreated floaters for examination, and ask the learners to observe and explain what might have occurred to account for these floaters.

Debriefing

When Working With Teachers

Discuss the quotes in light of some students' claims that they are just too dense to get science. The phenomenon of falling or sinking versus rising or floating can be considered a *visual participatory analogy* for the failure or success of students in learning science. Students' perceived learning deficiencies often are the result of curriculum-instruction-assessment that covers too much in too little time without adequate experiential and conceptual scaffolding to support the weight of so many new, heavy concepts. A student's ability to sink (fail) or rise (succeed) is always *relative* to the "buoyant force" provided by an intelligent CIA plan that activates the student's attention and catalyzes cognitive processing. Student sinkers can become

Dual-Density Discrepancies

risers who succeed in learning science if the CIA is enriched and differentiated to account for differences in prior knowledge and motivation, carefully designed multimedia are used to reduce the cognitive load (Mayer 2009), and more timely feedback is provided.

It is easy for teachers to forget that most of the basic core ideas in science took hundreds of years (and the work of exceptionally bright scientists such as Archimedes) to evolve to their current states. Teaching as telling and learning as listening are not sufficient to help students leap the gap between their preinstructional conceptions and modern scientific explanations. Interested teachers may explore the research on student misconceptions about density—floating or sinking, phases of matter, and physical changes (e.g., see Driver et al. 1994, chapters 8 and 9 as starting points)—to get a better sense of this challenge. Also, resource books and websites provide numerous concrete, real-world examples and applications for the vast majority of concepts taught at the K–12 level. With continued experiences with scientific modes of reasoning, students can construct not only a solid conceptual understanding of core ideas but also an increased ability to learn on their own—thereby greatly increasing their self-efficacy and potential interest in science and science careers. See Internet Connections: Conceptual Change Teaching.

When Working With Students

These experiments emphasize that floating or sinking depends on the *relative* density of the solid and the liquid (or two fluids) in question. As such, they can be presented as Engage-phase activities (without prematurely rushing to a complete explanation) that create a need to know that leads to quantitative mass and volume measurements in Explore-phase laboratory experiments (e.g., compare and contrast the volumes of the solid and liquid forms of water and butter). Alternatively, the experiments could be used as either teacher-directed Explain-phase activities or a form of curriculum-embedded formative assessments in the Elaborate-phase activities to assess whether students really get the idea of relative density as it applies to sinking and floating. As with other demonstrations done with magic or subterfuge, these experiments also present opportunities to work on students' skills of empirical observation, logical argumentation, and skeptical review as related to the nature of science and how we know what we know in science. Be sure to point out that ice is unusual in that it floats in pure water, whereas most substances are most dense in their solid phase and therefore their solid version will sink in their liquid version. This is one of a number of unique (for its low molecular mass) properties exhibited by water that are explained by hydrogen bonding. In this case, hydrogen bonding in crystalline water causes crystallized water to expand and become less dense than liquid water. The ice crystals also form an open structure that traps some of the previously dissolved air (but even ice cubes made from degassed water will float). See Internet Connection: PhET for a molecular animation of the three states of matter.

Extensions

A large number of variations exist on density demonstrations (e.g., see *Brain-Powered Science*, Activities #3, #4, #14, #31, and #32), including temperature-related convection cells and demonstrations using different concentrations of salt or sugar water. Examples of the latter include the following:

1. *Egg-citing Eggs-periment*: Fill three large, transparent cylinders (e.g., four-ball tennis cans) with, respectively, tap water, saturated salt water (~ 36 g NaCl/100 ml water; d = 1.2 g/ml), and a bilayer mixture that contains saturated salt water on the bottom and a layer of tap water on top that has been carefully poured down the side of the cylinder to fill it with minimum mixing. *Notes:*
    a. Kosher salt lacks hygroscopic additives (which are added to table salt to prevent it from absorbing water and clumping together) and works best to form a clear solution; you also can filter a solution of regular table salt so that it is clear like tap water.
    b. If they are not physically mixed, the two different-density liquids will stay layered and separate for quite some time before diffusion slowly mixes them.
    c. Either a fresh or hard-boiled egg (the latter's density = 1.09 g/cm^3) will sink in tap water, float in saturated salt water, and float at the nearly invisible interface where the fresh salt water

Dual-Density Discrepancies

begins in the third cylinder. Alternatively, standard-size potato cubes (d = 1.6 g/cm^3) or saturated sugar water (d = 1.8 g/ml) can be used (see Internet Connections). Use a similar questioning style as in the "Ice Is Nice" demonstration or have students measure the mass and volume of the solids (by water displacement) and liquids and calculate their densities. If desired, after the initial demonstration has been analyzed, a drop of food coloring placed and gently stirred on the top freshwater layer of the bilayer cylinder will make this layer visibly distinct from the salt water. Similar demonstration-experiments are described in the Internet Connections: Steve Spangler Science (Amazing Egg Experiments: The Rising Egg) and Becker Demonstrations (#25 Densi-Tee). The second source describes a multiday or multiweek display of a floating egg and golf ball (d = 1.15 g/ml) in a slowly diffusing column of freshwater layered on top of salt water. The process of density-related separation and mixing of freshwater and salt water has important biological implications in estuaries. The different densities of freshwater and salt water also affect the carrying capacity of oil tankers and other boats.

Another variation on the floating egg demonstration that involves chemical reactions is to place a fresh egg in a cylinder of hydrochloric acid. As the acid reacts with the calcium carbonate, carbon dioxide bubbles are produced that adhere to the shell and buoy up the egg and allow it to remain floating (similar to Activity #31 in *Brain-Powered Science*).

> **Safety Note**
> Indirectly vented chemical splash goggles are required.

2. *Liquid Rainbow Density Column*: Prepare a set of increasingly concentrated (and dense) saltwater (or sugar water) solutions in a series of test tubes colored with different food dyes. If a clear plastic straw (or glass tube) is submerged partway into the freshwater tube and an index finger is placed over the open end, one can withdraw a small amount of liquid. If this straw is then inserted into the next densest solution and the index finger-cover is released, the second liquid will enter at the bottom of the straw. If this process is repeated by pushing the straw a little deeper into each progressively denser liquid, students can easily get between three and five different-color layers in the plastic straw (with the

least dense on top, down to the most dense on the bottom). The Whelmers #64 website (Internet Connections) describes how to set this up as an inquiry-oriented hands-on exploration. Or the teacher can make a large-scale density column by carefully pouring solutions (from most to least dense) down the side of a large glass or plastic cylinder. See Becker #26 (Sucrose Density) and Steve Spangler (Seven-Layer Density Column) in the Internet Connections for this variation.

3. *Coke Floats ... or Does It?* A can of diet soda will float in room-temperature water (unless its volume has been reduced by denting it) while a can of regular soda will sink (see Wake Forest University Physics Department website in Internet Connections). Students can measure the density of each liquid and find or construct objects that will float in one but not the other liquid (e.g., construct a boat whose mass-to-volume ratio allows it to float only in regular Coke but sink in the lower-density Diet Coke). Alternatively, students can either test how much sugar or salt needs to be added to water to cause a can of regular soda to float or empirically determine what concentration of sugar is needed to match the soda's density. The mass difference between the two equal-volume cans may be dramatically visualized by placing a can of each type of soda on opposite sides of a double pan balance and adding packets of sugar on the diet soda side until it just counterbalances the mass of the can of regular soda. See Extension #4 for why this is a "weighty" matter. Also note that like other density experiments, this one can be readily explained in terms of the kinetic molecular theory and the idea of differentially sized moving molecules and "holes."

4. *"Too Much of Two Good Things"*: STS Case Studies: A 12 oz., 140 cal. can of regular soda contains the equivalent of about 10 tsp of sugar, and drinking one can per day could result in an annual weight gain of as much as 15 lbs. if the "extra" calories were not otherwise burned off! This fact can lead to an investigation of "empty calories" (i.e., foods that lack the nutritional value of vitamins, minerals, and protein) and the role of sugared soft drinks in the rise of obesity, dental problems, diabetes, and other health-related concerns in our society. Where the federal government supports artificially low-cost, corn-based sweeteners

via farm subsidies, a number of states have proposals for a "fat tax" on nondiet sugared soda in an effort to reduce consumption, fight obesity, and raise revenues. See Internet Connections: PhET for a related simulation on eating and exercise. In a similar vein, the health effects of excess sodium in American diets (i.e., an average of 3,400 mg/per day versus the 2,400 mg recommended maximum) is estimated to be a key contributing factor in more than 100,000 premature deaths annually; see Center for Science in the Public Interest and National Academy of Sciences reports to explore both issues.

5. *Hydrometer Hijinks*: A hydrometer is an instrument used to measure the relative density of a liquid as compared to water (or a liquid's specific gravity). Concentrations of various commercially important solutions are determined based on their specific gravity (e.g., milk, alcoholic beverages, petroleum products, lead-acid automobile batteries, automobile antifreeze solution, maple syrup). Students can explore how to make and calibrate their own hydrometer or commercial products designed for specific applications. See Wikipedia (Internet Connections).

Safety Note
Old hydrometers that contain mercury should never be used.

Internet Connections

- Becker Demonstrations and Experiments (Quick Time movies):

    #26 Sucrose Density (large, multicolored sugar water column): *http://chemmovies.unl.edu/chemistry/beckerdemos/BD026.html*

    #25: Densi-Tee (a golf ball in a water solution with gradually increasing salt concentration): *http://chemmovies.unl.edu/chemistry/beckerdemos/BD025.html*

- Center for Science in the Public Interest (food and nutrition watchdog): *www.cspinet.org*

- Conceptual Change Teaching (overviews of theory and research from various sources):

    www.physics.ohio-state.edu/~jossem/ICPE/C5.html

    www.ericdigests.org/2004-3/change.html

    http://narst.org/publications/research/concept.cfm

Activity 3

http://projects.coe.uga.edu/epltt/index.php?title=Conceptual_Change

www.agpa.uakron.edu/p16/btp.php?id=teaching-conceptual (contains a number of links)

- Disney Educational Productions: *Bill Nye the Science Guy: Buoyancy and Fluids* ($29.99/26 min. DVD): *http://dep.disney.go.com*

- Doing Chemistry (movie demonstrations): Density of Methanol versus Water (in a U-tube): *http://chemmovies.unl.edu/chemistry/dochem/DoChem002.html*

- HyperPhysics, Department of Physics & Astronomy, Georgia State University: Fluids: Select "Buoyancy" and "Archimedes Principle": *http://hyperphysics.phy-astr.gsu.edu/hbase/hframe.html*

- Java Applets on Physics: Select "Buoyant Force in Liquids": *www.walter-fendt.de/ph14e/index.html*

- National Center for Case Study Teaching in Science: A Cool Glass of Water: A Mystery (Will an ice cube melt faster in freshwater or salt water? Why or why not?): *www.sciencecases.org/melting_ice/melting_ice.asp*

- New-Skin Products: *www.newskinproducts.com/liquid-bandage.htm*

    See also Liquid Bandage: *http://en.wikipedia.org/wiki/Liquid_Bandage*

- PhET Interactive Simulations:
  States of Matter and Phase Changes (molecular animations):

    http://phet.colorado.edu/simulations/sims.php?sim=States_of_Matter

    Density (floating and sinking): *http://phet.colorado.edu/en/simulation/density*

    Eating and Exercise: *http://phet.colorado.edu/simulations/sims.php?sim=Eating_and_Exercise*

- SIMetric.Co.UK: Density tables: *www.simetric.co.uk/si_materials.htm*

- Steve Spangler Science Experiments: Density: See "Floating Lemons and Sinking Limes," "Amazing Egg Experiments," and "Seven-Layer Density Column": *www.stevespanglerscience.com/experiments/density-*

Dual-Density Discrepancies

- University of Iowa Physics and Astronomy Lecture Demonstrations (video clips): Heat and Fluids: *http://faraday.physics.uiowa.edu* (five demonstrations on density and buoyancy)
- University of Virginia Physics Department: Density HOEs with alcohol, water, salt, and ice plus graphing:

    http://galileo.phys.virginia.edu/outreach/8thGradeSOL/ InvestigatDensityFrm.htm

    http://galileo.phys.virginia.edu/outreach/8thGradeSOL/ LayeringDensityFrm.htm

    http://galileo.phys.virginia.edu/outreach/8thGradeSOL/ DensityLiquidFrm.htm

- Wake Forest University Physics Department: Physics of Matter videos: *www.wfu.edu/physics/demolabs/demos/avimov/bychptr/chptr4_matter.htm*
- Whelmers #49: Potato Float (in center of column): *www.mcrel.org/whelmers/whelm49.asp*
- Whelmers #64: Liquid Rainbow (layer five liquids): *www.mcrel.org/whelmers/whelm64.asp*
- Wikipedia: *http://en.wikipedia.org/wiki*. Search topics: buoyancy, density, estuary, and hydrometer

Answers to Questions in Procedure

Demonstration Experiment #1: Ice Is Nice, But Think Twice and It's Not My Fault, It's Salt!

7. No. A solid object will float in a liquid if it is less dense than the liquid and sink if it is more dense. Floating or sinking depends on the relative densities of the two substances involved. This principle applies equally to all phases of matter and therefore has many applications in the Earth's lithosphere (geology and plate tectonics), hydrosphere (freshwater hydrology and oceanography), and atmosphere (meteorology). The densities (in g/ml or g/cm^3) of the substances used in this experiment are halite (sodium choride/2.16), liquid water (1.00), ice (0.92), and ethanol (0.79). If desired, a piece of paraffin wax (d = 0.85) can be

observed to float in water but sink in ethanol (and methanol or isopropanol). Also, an ice cube will sink in corn oil (0.91) and float in corn syrup (1.38).

Demonstration Experiment #2: Sugar Is Sweet, But When It Floats, It's Neat

2. Most learners will predict that the sugar cubes will sink and dissolve in water. Dissolving is considered to be a physical change because chemical bonds are not broken and reformed to form new chemical substances, and the original sugar crystals are readily reformed when the water is allowed to evaporate off or is boiled away. Chemical reactions often involve color changes, the evolution of a gas, or the formation of a precipitate (when two liquids are mixed together). However, colors can change as a result of a simple physical mixing of pigments (e.g., food coloring) that can be readily separated by chromatography; crystals can form from a saturated solution of the same chemical; and boiling liquids give off gases. These are all examples of physical changes. Some solids will both dissolve and chemically react when they are placed in water or other liquids. For example, most metals react in acid solutions (e.g., zinc added to hydrochloric acid generates hydrogen gas, and copper placed in nitric acid generates the toxic orange-brown-colored nitrogen dioxide and a green copper nitrate solution) and very chemically reactive metals (Group I alkali) will even explode when placed in water.

3. Increasing the temperature of the water increases the average kinetic energy of the molecules and causes them to hit the surface of the cube with more force and greater frequency. Also, breaking up the sugar cubes into smaller pieces increases their surface-area-to-volume ratio, and stirring the solution makes it easier for water molecules to get to the sugar crystals.

4. The sugar cubes are hollowed-out frames rather than solid. The pyroxylin (or cellulose nitrate or collodion cotton) in New-Skin forms a lattice-like, semipermeable membrane that allows water to pass into the sugar cube and sugar water to diffuse out. The collodion framework or shell that remains is buoyant enough to float. Display New-Skin and discuss its use.

Inferences, Inquiry, and Insight
Meaningful "Miss-takes"

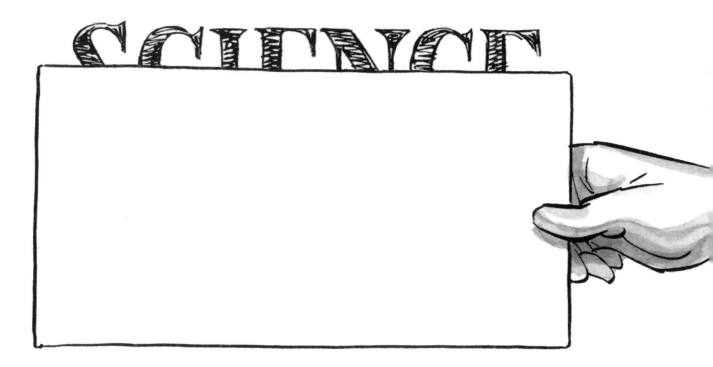

Expected Outcome

Learners are surprised to find that an obvious, logical inference about a partially hidden word pattern proves to be incorrect—the pattern does not spell out the word *SCIENCE*. Instead, when the lower half of the pattern is revealed, it is actually a sequence of letters and numbers rather than letters only (SGI5NG5).

Science Concepts

The Nature of Science (NOS) includes making scientific inferences based on available empirical evidence (observation) and logical arguments (e.g., prediction based on pattern recognition). Making and learning from "miss-takes" uncovered via skeptical review is an essential component of both individual learning and tentative, subject-to-revision scientific explanations. Scientific progress depends on our ability to make, test, and refine inferential models and to develop new technologies to peer inside complex black box systems.

Science Education Concepts

The *Benchmarks for Science Literacy* (AAAS 1993), the *National Science Education Standards* (NRC 1996, 2000) and the *Framework for Science Education, Preliminary Public Draft* (NRC 2010) emphasize the importance of teaching students about the NOS through inquiry-oriented activities that allow students to gather data, make inferences, and critically review their empirical results to modify their initial theories as necessary. This book and its predecessor contain many hands-on explorations (HOEs) designed to achieve this goal. But quick, no-cost paper-and-pencil puzzles (PPPs) can also serve as *visual participatory analogies* for and models of minds-on inquiry into black box systems that are temporarily unopenable.

This discrepant-event activity also demonstrates the pedagogical principle of the power of *puzzles and discrepant or counterintuitive events to activate attention and catalyze cognitive processing* within learners (Baddock and Bucat 2008; Bell 2008; Chinn and Brewer 1993, 1998; O'Brien 2010; Posner et al. 1982). See also Activities #15 and #16 in *Brain-Powered Science*, which emphasize and discuss this same principle. Natural selection (survival of the fittest) of our species over evolutionary history has genetically prewired the human brain to look for patterns, notice anomalies, and try to resolve puzzling phenomena. Effective teaching draws on this innate curiosity and puzzle-solving propensity. Though excessive stress can restrict our ability to learn, eustress (or good stress; see Internet Connections: Eustress) optimally challenges learners' current knowledge and abilities and expands their sense of self-efficacy. Too little challenge produces boredom; too much challenge results in frustration and anxiety (and, over time, learned helplessness). Optimizing the cognitive load by working within and constantly stretching each individual learner's zone of proximal development (i.e., the Goldilocks pedagogical principle) are two of the key challenges of teaching diverse learners.

Materials

- Transparency of the "word" *SGI5NG5* in one of the fonts listed in Procedure, step #1
- 5 in. × 8 in. index card or sticky note to cover the top half of this "word"
- Overhead projector or document camera to project an image

Points to Ponder

The moment a person forms a theory, his imagination sees, in every object, only the traits that favor that theory.
—Thomas Jefferson, American president and amateur scientist (1743–1826)

The world little knows how many thoughts and theories which have passed through the mind of a scientific investigator and have been crushed in the silence and secrecy by his own severe criticism and adverse examinations; that in the most successful instances not a tenth of the suggestions, the hopes, the wishes, the preliminary conclusions have been realized.
—Michael Faraday, English chemist and physicist (1791–1867)

The great tragedy of Science—the slaying of a beautiful hypothesis by an ugly fact.
—Thomas Henry Huxley, English biologist and evolutionist (1825–1895)

… progress in science comes when experiments contradict theory.
—Richard Feynman, American physicist (1918–1988)

Procedure

1. Display on an overhead projector or document camera the top half of what appears to be the word *SCIENCE*, but is in fact *SGI5NG5*, using one of the fonts listed below (type size 72 or greater). In these fonts, the top half of the number *5* looks like the top half of the block letter *E* and the top half of the letter *G* looks like the top half of the letter *C*. You also may want to include (but cover) the word *SCIENCE* in the same font size and type just below the discrepant nonword. Ask: How is this puzzle analogous to the process of science?

    SGI5NG5 (Helvetica)   SGI5NG5 (Courier)
    SGI5NG5 (Tahoma)   SGI5NG5 (Verdana)

2. Ask the learners to consider this puzzle a *visual participatory analogy* for cases in science in which we have only part of the data in hand. Challenge students to think-write-pair share specific examples where scientists try to complete a jigsaw puzzle when they have only some of the pieces and must therefore make *inferences* about what's missing and its order relative to the data they have in hand. Then challenge students to make an educated guess or predict what the remainder of this particular puzzle might reveal based on *extrapolations* from the available *empirical evidence*. After allowing dyads to write their ideas down, ask them to share with the class as a whole. If the learners do not generate multiple working hypotheses (i.e., they all prematurely settle on "science" as their answer), challenge them to consider alternatives.

3. Remove the 5 in. × 8 in. index card or sticky notes and ask the learners to observe the actual message. Ask: When viewed as an analogy for the process of science, what does the act of removing the card represent?

4. Discuss and explain how extrapolation from available data can sometimes be incorrect, how many scientific phenomena can be considered to be, at least in part, black box systems, and why science moves forward by *empirical evidence* (observations), *logical argument*, and *skeptical review* (which includes further testing). It is the nature of science to remain forever open to further insights

Inferences, Inquiry, and Insight

and modification. Note that in this particular case any number of possible answers could be right, including (a) the word *science* and (b) having no additional information below the top half at all. Results that run counter to our working hypotheses often are the origin of scientific breakthroughs (see Internet Connections: Serendipity in Science). Scientific progress depends on the synergy between creative (or speculative) and critical (or skeptical) thinking.

Debriefing

When Working With Teachers

Discuss the pedagogical power of puzzles and discrepant or counterintuitive events to capture students' attention and motivate their active minds-on cognitive processing. Letting students make "miss-takes" and experience unexpected or discrepant results helps them learn both the nature of science and how to learn (and intentionally unlearn outdated) science. Discuss how the quotes on page 49 relate to the need for constructivist teaching approaches in which the teacher helps students reconstruct prior understanding and assumptions in light of new and sometimes discrepant experiences that may contradict what they believe to be so. From a Piagetian perspective, new information can sometimes be directly assimilated into pre-existing conceptual schema, but sometimes accommodation or modification of past schema is necessary for the new anomalous information to fit in and make sense. The history of science includes many cases where once well-received theories or scientific paradigms were displaced by better, more encompassing theories (e.g., phlogiston theory in chemistry, caloric theory in physics, spontaneous generation in biology, and geocentric theory in Earth and space science). Teachers may wish to explore classic examples of theories that eventually failed:

- Grant, J. 2006. *Discarded science: Ideas that seemed good at the time.* Wisley, Surrey, UK: Facts, Figures and Fun. (An amusing tour of many once scientifically respectable but now untenable hypotheses, such as the flat Earth and Piltdown Man)
- Youngson, R. 1998. *Scientific blunders: A brief history of how wrong scientists can sometimes be.* New York: Carroll & Graf Publishers. (Contains numerous examples of scientific "miss-takes" from the early history of science up until cold fusion)

When Working With Students

Discuss how prior cognitive commitments, conservatism, and inertia often make it difficult for nonscientists and scientists alike to modify or replace prior beliefs in light of new information and how learning science involves modifying or replacing some of what we "know" that isn't so. Share and discuss the quotes on page 49 in terms of how they relate to the nature of science (i.e., evidence, models, and explanations) and the challenge of learning science. Analogically, science can be visualized as a magnificent edifice that is constantly under reconstruction, remodeling, and renovation by individuals and teams that are both international and intergenerational. So, too, individual human conceptual understanding is always a work under construction and in progress.

Extensions

1. *Reverse Pattern Finding*: Challenge learners to decipher the meaning of the phrase *MX GHR QE TFA*. This can be considered the reverse of the *SGI5NG5* problem in that if the lower half of the puzzle is covered, most readers will "see" (readily infer) the phrase *MY CUP OF TEA*. This wordplay puzzle is attributed to Stefan Van den Bergh at the Planet Perplex website (see Internet Connections). Learners may enjoy seeing if they can create similar phrases where the human brain will infer missing information and see a pattern or phrase that may (or may not) actually be there. Suggest that science really could be all students' "cup of tea" if presented in a way that was FUNdaMENTAL, or both fun work and hard play! Effective teachers develop students' abilities to retain new information and the dispositions and abilities to creatively transfer ideas and skills to new contexts.

2. *FUNdaMENTALs of Science With Plexars, Pundles, Frame Games, and Wordies*: These visual-verbal puzzles use the relative location, orientation, size, shape, or sound of letters, words, numbers, or punctuation marks to spell out a hidden word or phrase, object, or event in a way that challenges our conventional sense of spelling and grammar. As such, they bend the rules of reading and encourage the development of a variety of divergent, creative, critical-thinking skills. These puzzles can be used in

Inferences, Inquiry, and Insight

science classes as quick preclass "neurobic" warm-ups or mental push-ups, motivational invitations to inquiry, transitions between topics, creative lesson closures linked to key concepts, or bonus point opportunities on tests. Depending on the context and purpose, they may be assigned to individual students, dyads, or cooperative learning teams via posting on the bulletin board, chalkboard or whiteboard, multimedia projector, or student handouts. Compilations of these types of puzzles can be found in the following resources:

- Halperin, J. 1979. *Symbol simons: A new type of word game and puzzle.* New York: Wanderer Books/Simon & Schuster. (189 word puzzles; see also *Symbol Simons Too* [1981])
- Hammond, D., T. Lester, and J. Scales. 1999. *Science plexers: A collection of word puzzles.* Parsippany, NJ: Dale Seymour Publications/Addison Wesley Longman. (More than 240 reproducible puzzles)
- Moog, B. 2004. *Armchair puzzlers: Symbol simon.* San Francisco, CA: University Games. (80 puzzles combine symbols with pictures to depict a particular item.)
- Nash, B., and G. Nash. 1979. *Pundles.* New York: Stonesong Press/Grosset & Dunlap. (90 puzzles)
- Price, R., R. A. Lovka, and B. Lovka. 2000. *Classic droodles.* Beverly Hills, CA: Tallfellow Press. (Created by Roger Price in the 1950s, a Droodle [which is a doodle plus a riddle] is a kind of minimalist sketch cartoon featuring somewhat abstract pictorial elements accompanied by the implicit question "What is it?" which requires inferential reasoning and creativity to create the appropriate caption.)
- Stickels, T. 2004. *Sit and solve frame games.* Sterling. 96-page book offers visual puns, almost like rebuses, with a combination of writing and images. Frame games appear in many newspapers (in the colored Sunday insert section) and challenge readers to come up with a well-known saying, person, place, or thing. See the website in Internet Connections.

Internet Connections
(for Visual-Verbal Puzzles)

- American Museum of Natural History: Brain: The Inside Story: *www.amnh.org/exhibitions/brain/index.php*
- Bills.Games.com: *www.billsgames.com/brain-teasers* (More than 260 pundles posted)
- *Brain Rules: 12 Principles of Surviving and Thriving at Work, Home and School*. Book, DVD, and website (by John Medina): *http://brainrules.net*
- Braingle Teasers, Riddles, Games, Forums, and more: *www.braingle.com*
- Cognitive load theory: *http://en.wikipedia.org/wiki/Cognitive_load_theory*
- Discovery (Channel) Education: PuzzleMaker: Criss-cross, word search, and other templates: *http://puzzlemaker.discoveryeducation.com*
- Disney Educational Productions: *Bill Nye the Science Guy: Patterns* ($29.99/26 min. DVD): *http://dep.disney.go.com*
- Droodles: Search Google
- Eustress: *http://en.wikipedia.org/wiki/Eustress*
- MakeBeliefsComix! Make your own comic strip generator: *www.makebeliefscomix.com*
- National Center for Case Study Teaching in Science: *http://sciencecases.lib.buffalo.edu/cs*
- Planet Perplex: MX GHR TFA puzzle: *http://planetperplex.com/en/item151*
- Serendipity in Science:

    Access Excellence: Discovery, Chance & the Scientific Method: *www.accessexcellence.org/AE/AEC/CC/chance.html*

    Wikipedia: (links to many examples): *http://en.wikipedia.org/wiki/Serendipity*

Inferences, Inquiry, and Insight

- *Sleights of Mind: What the Neuroscience of Magic Reveals About Our Everyday Deceptions* (book by S.L. Macknik and Susana Martinez-Conde; includes multimedia clips): *www.sleightsofmind.com*
- Society for Neuroscience: *Brain Facts: A Primer on the Brain and Nervous System:* 74-page book, CD, and free PDF: *www.sfn.org/index.aspx?pagename=brainfacts*
- Terry Stickels: Puzzles, Frame Games & Gallery: *www.terrystickels.com/puzzles.html*
- Wordies on the Web: *www.wordies.ca* (A sampling of puzzles from a total of 900 found in the two volumes of the 52-page books *In Other Words: A Collection of Verbal Picture Puzzles, Book One and Book Two*. Access books through the website.)
- Zone of Proximal Development: *http://en.wikipedia.org/wiki/Zone_of_Proximal_Development*

Answers to Questions in Procedure, steps #1–#3

1. The puzzle is analogous to a scientist having some empirical evidence in hand, and, lacking some relevant data, makes an inference that makes sense of the data in terms of a bigger picture theory.

2. The fields of paleontology, geology, forensic sciences, meteorology and climatology, and, to some extent, all sciences are full of examples in which we have only part of the data or some of the data may be erroneous, and we need to fill in missing pieces by making informed inferences. In the case of this visual puzzle, a hypothesis that makes sense is that the missing data would complete the word *science*. It is important to note that science is based on the assumption that nature does indeed make sense in light of universal laws and theories that govern cause-and-effect relationships and create meaningful, repeating patterns.

3. Removing the card is analogous to uncovering or discovering additional empirical evidence as a result of additional experimentation with improved methodologies and refined questions.

Pseudoscience in the News
Preposterous Propositions
and Media Mayhem Matters

Expected Outcome

Pseudoscience "news" is critiqued to help identify the defining attributes of science. An Extension activity features a critical review of the legitimacy of astrology.

Science Concepts

The nature of science (NOS) is to rely on empirical evidence, logical argument, and skeptical review, versus pseudoscience (e.g., astrology), which falsely claims scientific validity and credibility despite the weight of contradictory evidence. Both science and pseudoscience are creative, but the latter is not balanced by critical, skeptical review by a community of scholars (or when it is, it does not hold up or has been falsified). People in the pseudosciences disseminate misconceptions and myths about the natural world that scientifically literate individuals need to challenge with scientifically sound arguments to avoid personal and public policy, science-technology-society (STS), or science-technology-engineering-mathematics (STEM) errors.

Science Education Concepts

Local newspapers devote varying amounts of space to science, science *fiction* (e.g., book and movie reviews), and *pseudo*science "news" stories (e.g., astrological horoscopes and "miracle medicine" ads). Informed readers have an intuitive sense of how these domains are different, although the lines can sometimes become blurred. Especially problematic are national news magazines, which typically publish "news stories" that shamelessly (for entertainment value) blend all three types of writing into single articles. Helping students become informed, scientifically literate consumers of the popular press is especially critical when dealing with such STS issues as health, diet and exercise, environmental and energy policies, and input into publicly funded research. Students' fascination with pseudoscience can be used to bring students to the truly "far out" world of real science.

This activity can be modeled with teachers fairly quickly but is best done (especially with grades 5–12 students) during one or more class periods. This "read, categorize, and discuss" type of discrepant-event activity also demonstrates the pedagogical principle of the power of *cognitive connections and meaningfulness in activating attention and catalyzing cognitive processing* (Donovan and Bransford 2005; Mintzes, Wandersee, and Novak 1998). That is, activities that connect with and build on students' prior interests and knowledge are perceived as making more sense than those that don't explicitly help students construct these kinds of connections. (Activities #17 and #18 in *Brain-Powered Science* also feature this central principle of constructivist learning theory.) However, human beings' innate, evolutionarily driven, genetically programmed search for meaningful patterns can sometimes lead us to mistakenly see patterns and cause-effect relationships where they do not exist. Skeptical review by individual researchers and the scientific community is essential to constantly guard against and correct such errors.

Materials

Mixed samples of articles on science and pseudoscience can be clipped from local newspapers, national tabloid newspapers (e.g., *Globe, National Enquirer,* and *The Sun*), and the internet, with the source citations removed (but coded by the teacher). Take care in the selection of appropriate stories and photos, given the intentionally scandalous nature of news magazines.

Procedure

1. Have the learners work in small groups to critically analyze and categorize the sample articles in terms of (a) their relative motivational appeal, understandability, and relevance to individual readers and society and (b) whether they seem to depict *science* or *pseudoscience*. Ask the groups to develop a list of criteria to distinguish these two domains. Share the quotes and ask the groups to consider how they relate to their criteria. In particular, how do the two domains deal with the ideas of intellectual parsimony and falsification?

2. After reviewing and categorizing the articles, discuss the following questions: Does it appear that science sells as well as the more sensational, pseudoscience domain? Why or why not? Based on your critical review, do you think media matter? What harm (if any) is there with pseudoscience?

3. If time permits, focus specifically on astrological horoscopes (see also Extension #1). Does astrology pass the test of being scientific? Why or why not? *Note:* Johannes Kepler (1571–1630) was the last great astronomer whose source of income included

Points to Ponder

Astrology is a sickness, not a science ... It is a tree under the shade of which all sorts of superstitions thrive ... Only fools and charlatans lend value to it.
—Moses ben Maimon (or Maimonides), Spanish-born Jewish religious philosopher and physician (1135–1204)

We are to admit no more causes of natural things than such are both true and sufficient to explain their appearance. Therefore, to the same natural effects we must, so far as possible, assign the same causes.
—Isaac Newton, English scientist and mathematician (1642–1727). This statement is a refinement of Occam's Razor, a key philosophical building block of science.

Science is simply common sense at its best—that is, rigidly accurate in observation, and merciless to fallacy in logic.
—Thomas Henry Huxley, English biologist and evolutionist (1825–1895)

The grand aim of science is ... to cover the greatest possible number of empirical facts by logical deductions from the smallest number of hypotheses or axioms.
—Albert Einstein, German-born American physicist (1879–1955)

The true scientific attitude is utterly different from the dogmatic attitude which constantly claimed to find "verification" for its favourite theories ... the scientific attitude was the critical attitude, which did not look for verifications but for crucial tests; tests which could refute the theory being tested, though they could never establish it.
—Karl Popper, Austrian philosopher of science (1902–1994)

Extraordinary claims require extraordinary proofs.
—Carl Sagan, American astronomer and exobiologist (1934–1996)

astrological horoscopes, a common practice in his day. Develop a list of questions that a skeptical scientist would use to challenge the validity of the advice and predictions in terms of the triple litmus tests of *empirical evidence, logical argument,* and *skeptical review*. Could your questions convince a true believer that astrology does not pass the test of being real science? Would human common sense be more of an aid or a hindrance in your debunking role?

Debriefing

When Working With Teachers

To quickly model this activity with teachers, review the triple litmus test for science and the meaning of the quotes on page 60. Initiate a brief discussion of the following open-ended questions (that can be continued online):

1. Is legitimate science writing typically as engaging as pseudoscience (or even science fiction)? How do the popular writings of modern-day scientist-authors (e.g., Isaac Asimov, Richard Feynman, Stephen Jay Gould, Stephen Hawking, Carl Sagan, Lewis Thomas, and E.O. Wilson) achieve the objective of being scientifically accurate, interesting, and educational?

2. What lessons could textbook authors and publishers learn from news magazines?

3. How can teachers creatively use both newspaper science and news magazine pseudoscience to teach media literacy and critical thinking?

4. How should a teacher's curriculum-instruction-assessment help students simultaneously develop both creative and critical-thinking skills?

When Working With Students

Unsubstantiated and unsupported pseudoscientific claims cover a wide range of topics, including alien abductions, antigravity devices, astrology, biofield therapeutics, cold fusion, creation science, dowsing, EMF-

induced cancer, homeopathy, new age cures, palm reading, parapsychology, perpetual motion machines, poltergeists, pyramid crystal power, and telekinesis. Also called junk, pathological and voodoo science, and superstitions, such STS topics are generally distinguishable from fringe or cutting-edge science that pushes the boundaries of well-known, established ideas (e.g., scientific studies of ancient therapeutic practices whose efficacy seems to be more than the placebo effect). Investigations related to NOS and "science" in the popular press can become a year-long activity. It can become a regular feature during the Engage and Evaluate phases of each unit, where students are challenged to help the teacher find and debunk pseudoscience claims related to the unit's content focus. The Discovery Channel show *MythBusters* can serve as both a model and a resource (see Internet Connections).

Extensions

1. Astrological Horoscopes and the Triple Test of Scientific Validity

The following questions for critical consideration serve as a resource for teachers who wish to expand on the activity for use with students.

Empirical Evidence

a. Do the predictions and advice for your astrological sign seem to ring true for you for a given specific day, or does another sign seem to be a better match or statistical correlate for the actual events of your life on that day? How could you test this in a way that personal belief and after-the-fact interpretations would be less likely to bias the results?

b. In examining the predictions and advice for your sign over a longer period (i.e., one week), do you see evidence of a good fit with the data of your life experiences? Could the lack of specificity (or vagueness) of the predictions be a factor?

c. Does the accuracy of horoscopes hold for a broad sampling of people across all twelve signs over a longer period of testing? Could our tendency to forget bad guesses and remember accurate ones affect your survey results (especially if the test subjects are true believers)?

Logical Arguments

d. How are astrological horoscopes different from both science and Chinese fortune cookies?

e. Assuming that the astrological horoscopes have predictive validity, what natural forces could potentially explain the observed influence of the planets and stars?

f. What do the inverse square law and the law of universal gravitation suggest about the relative influence of moderately large nearby objects (natural or man-made) as compared to even much larger but much farther-away objects such as planets and stars?

g. Are known natural forces (i.e., electromagnetic and gravitational) that act on a fetus or newborn substantially different before birth versus after?

h. What does the relatively modern discovery of the outer planets of Uranus (1781), Neptune (1846), and Pluto (1930) suggest about the validity and reliability of horoscopes made by the ancient astrologers and early modern astronomer-astrologists (e.g., Copernicus, Brahe, and Kepler)? Should horoscopes made today be much more precise and accurate than older ones?

Skeptical Review

i. How could reading your horoscope *before* you start your day influence the way you interpret and respond to particular events (i.e., could it become a self-fulfilling prophecy)? What is the value of an independent observer and repeated, multiple, double-blind studies in supporting scientific theories? Does astrology stand the light of day on these kinds of tests?

j. Would a child's future be altered if it was delivered by a Cesarean section a day before it was naturally due and if the premature delivery occurred on the day before a change in sign?

k. Would identical twins born on subsequent days be likely to have very different personalities and life experiences based on their respective signs? (*Note:* This somewhat infrequent event does occasionally occur even at the December 31–January 1 New Year's break.) Could a one-second difference between two births really be expected to cause significant differences in personalities and life experiences?

l. Have the dates of the various signs been updated from ancient times to modern times with changes in the calendar? What happens when a given star goes supernova and no longer appears in the night sky? (*Note:* There can be a very long gap between a star's actual "disappearance" and our noticing the event on Earth; when we look up at stars, we are always viewing the distant past, not the current reality.)

m. Do Western, Greco-Roman signs and predictions correlate with the Chinese calendar's corresponding animal-designated years, Zodiac signs, and horoscopes? Are valid scientific concepts and theories universals or are they culture-dependent?

n. If natural forces account for the influence of planets and stars, should geographical location (as partially defined by human history and culture) be a significant variable in addition to the birth date?

o. If the relative positions of the various planets and stars on the day you were born affected your future, shouldn't all changes be fine gradations rather than 12 distinctly different types?

p. What do you think Shakespeare meant in his play *Julius Caesar* when he wrote, "The fault, dear Brutus, is not in our stars, / But in ourselves"?

q. Is astro<u>nomy</u> or astro<u>logy</u> more akin to bio<u>logy</u> and geo<u>logy</u>? Why? Do you think a certain amount of "street credibility" comes with a name? Astronomy, like all true sciences, is intimately connected with the sciences of chemistry, physics, and mathematics (especially calculus). What does the lack of such connections suggest about astrology?

2. Science, Pseudoscience, Critical Thinking, and Skepticism

Challenge students to use their critical-thinking skills and scientific skepticism to (a) compare and contrast various reputable science news websites with those that feature pseudoscience, scientific hoaxes, myths, and urban legends, or (b) critically review the science-to-pseudoscience ratio in popular movies (see Blick on Flicks and other Internet Connections). See also Activities #2 and #22 for other references.

Pseudoscience in the News

Internet Connections

- AFU and Urban Legends Archive: *http://tafkac.org*
- Alien Autopsy (special effects debunking): Use Google or YouTube.
- Astrology.com (horoscopes): *www.astrology.com*
- Astronomical Society of the Pacific: *www.astrosociety.org*
- *Bad Astronomy* (author Phil Plait's site on misconceptions and more): *www.badastronomy.com/index.html*
- Biology Corner (go to Debunking the Paranormal: Evidence of Yeti activity with rubric): *www.biologycorner.com/lesson-plans/scientific-method*
- Blick on Flicks feature of NSTA Reports: *www.nsta.org/blickonflicks*
- Carl Sagan's Baloney Detection Kit (based on his book *The Demon Haunted World*):

    www.carlsagan.com

    http://skepticreport.com/sr/?p=152

    http://users.tpg.com.au/users/tps-seti/baloney.html

- Committee for Scientific Investigations of Claims of the Paranormal (*Skeptical Inquirer*): *www.csicop.org* (Variety of resources, including *Field Guide to Critical Thinking: www.csicop.org/si/show/field_guide_to_critical_thinking*)
- Disney Educational Productions: *Bill Nye the Science Guy: Pseudoscience* ($29.99/26 min. DVD; see also the *Eye of Nye* episode on same topic for grades 7–12): *http://dep.disney.go.com*
- DMOZ (links to paranormal topics): *www.dmoz.org/Society/Paranormal*
- *Enemies of Reason: Part 1: Astrology:* Richard Dawkins' debunking investigation videoclip:

    http://onegoodmovemedia.org/movies/0708/enemiesofreason_astrology.mov

    http://topdocumentaryfilms.com/enemies-reason

- *Heavenly Errors* (author Neil F. Comins's site on astronomy-related misconceptions): *www.physics.umaine.edu/ncomins*

Activity 5

- Insultingly Stupid Movie Physics: *http://intuitor.com/moviephysics* (see also book)
- James Randi (magician and pseudoscience debunker) Educational Foundation: *www.randi.org*
- Junk Science Home Page (pseudoscience in news): *www.junkscience.com*
- Museum of Hoaxes: *www.museumofhoaxes.com*
- *MythBusters*: *http://dsc.discovery.com/fansites/mythbusters/about/about.html*
- National Center for Case Study Teaching in Science at the SUNY-Buffalo:

    Extrasensory Perception: Pseudoscience? A Battle at the Edge of Science: www.sciencecases.org/esp/esp.asp

    The "Mozart Effect": Psychological Research Methods: *www.sciencecases.org/psych_research/psych_research.asp*

- Natural History Museum: *The Case of the Piltdown Man*: *www.nhm.ac.uk/nature-online/life/human-origins/piltdown-man*
- NOVA: *Secrets of the Psychics* (video debunking of pseudoscience by the Amazing Randi): *www.pbs.org/wgbh/nova/teachers/programs/2012_psychics.html*
- PBS Teachers Resource Roundups: Developing Scientific Thinking (PDF download): *www.pbs.org/teachers/resourceroundups*
- Project Look Sharp Media Literacy: See Media Literacy Handouts, Curriculum Kits, and Research and Assessment: *www.ithaca.edu/looksharp*
- QuackWatch (debunks medical fraud): *www.quackwatch.com*
- Richard Dawkins Foundation for Reason and Science: *http://richarddawkins.net*
- Science Hobbyist Weird Science: *www.amasci.com/weird.html*
- Science Humor—It's Alive (video clip spoofs of pseudoscience): *www.sciencehumor.org*
- Science News on the Internet (reputable sites by legitimate scientific sources):

    BBC Science and Nature: *www.bbc.co.uk/sn*

Pseudoscience in the News

CNN: *www.cnn.com/TECH/science/archive/index.html*

EurekAlert!: *www.eurekalert.org* (sponsored by AAAS)

National Public Radio: *www.npr.org/sections/science*

New York Times: *www.nytimes.com/pages/science/index.html*

Science Daily: *www.sciencedaily.com*

Science Friday: *www.sciencefriday.com*

- *Scientific American Frontiers* Show 802: Beyond Science? (November 1997): *www.pbs.org/safarchive/4_class/45_pguides/pguide_1005/44105_catalog.html*
- Seven Warning Signs of Bogus Science (article by Robert L. Park, *Chronicle of Higher Education*, 2003): *http://chronicle.com/free/v49/i21/21b02001.htm*
- The Skeptic's Dictionary: *www.skepdic.com*. See pseudoscience and other entries.
- The Skeptics Society (A Skeptical Manifesto by Michael Shermer): *http://skeptic.com*
- *Sleights of Mind: What the Neuroscience of Magic Reveals About Our Everyday Deceptions* (book by S.L. Macknik and Susana Martinez-Conde; includes multimedia clips): *www.sleightsofmind.com*
- Snopes.com Urban Legends References Pages: *www.snopes.com/snopes.asp*
- TruthorFiction.com (reality check on internet and e-mail stories): *www.truthorfiction.com*
- Wikipedia: *http://en.wikipedia.org/wiki*. Search topics: astrology (versus astronomy), cartoon physics, constructivism (learning theory), hoaxes in science, Karl Popper, Occam's Razor, and pseudoscience.

Answers to Questions in Procedure, steps #1–#3

1. Many learners will probably find pseudoscience articles (and science fiction stories) to be more engaging than science articles. This is due to both the style of writing and the emotional, wishful-thinking

orientation of the former relative to the latter. However, some learners will be more naturally skeptical and find the lack of evidence and logic to be a turnoff. Once the quotes on page 60 are shared, the contrasts should be evident to all. Both domains use overarching theories or big ideas to attempt to explain a large number of observations, but only science invites and expects its theories to be progressively challenged, improved, and replaced based on empirical tests that have the potential to falsify its basic tenets. Pseudoscience does not seek out or attend to the results of critical tests that can (and have) falsified its claims of scientific validity.

2. People readily believe what they wish to be true, especially when it comes to quick-fix, low-cost "miracle cures" related to health, diet, beauty, energy and environmental problems, and other STS-related issues. Pseudoscience solutions may be directly harmful, or at least they can distract us from doing the hard work necessary to find and implement real solutions. Ignorance is *not* bliss.

3. Although astronomy and astrology were indistinguishable prior to the Renaissance, the latter does not pass the test of being scientific (see Extension #1). Unfortunately, empirical evidence, logical argument, and skeptical reasoning are necessary but may be insufficient for debunking pseudoscience in the minds of true believers. Neither common sense nor scientific habits of mind are as common as one would like. Surveys consistently indicate that a large percentage of Americans believe in astrology and other pseudoscience domains despite the fact that most have been consistently falsified.

Debriefing

When Working With Teachers

1. Fiction writers have far fewer constraints on their creativity than do science fact writers. That said, science popularizers are known for their clever use of analogies, their ability to use words to paint pictures, and their ability to connect with and expand their readers' prior knowledge, interests, and emotional commitments. Science writing (including textbooks) need not be dry, lifeless, and boring!

2. Like the internet, textbooks can be considered virtual teachers for students to the extent they activate the readers' attention and catalyze their minds-on cognitive processing of big, important, and compelling ideas. The development of science standards, intra- and inter-grade-level learning progressions, and research-informed and technology-enhanced textbooks promises to change student passivity and lack of interest.

3. See in the Internet Connections the Project Look Sharp Media Literacy website as a sample resource for teaching media literacy and critical thinking.

4. Consider the analogy of learning as an act of construction in which the teacher's curriculum-instruction-assessment plan is the blueprint that the student-owner uses to build and remodel his or her own cognitive structure-house with assistance from the teacher-general contractor.

Extensions

a.–c. A variety of means could be used to test the validity of horoscopes while controlling for human gullibility. Examples include using blind match tests to see if test subjects are as likely to see mismatched horoscopes as being true for them as correctly matched ones, having test subjects write their own after-the-fact horoscope for a given day and then compare it to their real horoscope for that day, and critically examining cases where students' zodiac signs and corresponding horoscopes differ by only a day or two. In all cases, using larger numbers of test subjects, longer test periods, statistical analysis, and independent raters would be helpful, but the vagueness of horoscopes may make it difficult to falsify their predictive power. The ability to be falsified in fair tests is a key attribute of scientific theories and hypotheses.

d.–h. Science develops theories that have both explanatory power and predictive validity to account for known observations and patterns and to forecast future ones. Fortune cookies are a "luck of the draw" event, whereas horoscopes are presumably linked to specific astronomical orientations on the day of a person's birth. Electromagnetic, gravitational, and strong and weak nuclear forces are the four fundamental forces known to exist. Given the scale of distances involved, the former two forces are potentially relevant to explain any presumed astrological effects. However, the inverse square laws that apply to electromagnetic radiation and gravity suggest that nearby objects of even modest weight (another human being in the delivery room) exert larger gravitational attractions on humans and create larger radiation gains (i.e., all humans contain a low level of naturally occurring radioisotopes) than very distant (i.e., beyond the Moon) astronomical objects. The Moon has a gravitational effect on the Earth (and humans) that is more than 50 times the combined gravitational force of all the planets in our solar system (because of its relative closeness). In any case, the electromagnetic and gravitational forces that act on a fetus or newborn are not substantially different before birth than they are afterwards. The fact that the three outer planets were not known until the modern era would suggest that today's horoscopes should be more precise and accurate than older ones. No evidence to support this claim exists.

i.–q. A critical assessment of the answers to any or all of these skeptical review questions indicates that the lack of evidence (or a preponderance of counterevidence) and logic in astrology explains why it is considered a pseudoscience. Note that the word *astronomy* is derived from the Greek words for *star* (*astron*) and *law* (*nomos*). Astronomy is intimately connected with the sciences of chemistry, physics, and mathematics (especially calculus), whereas astrology lacks such interdisciplinary connections. Valid scientific fields never stand in complete isolation.

Activity 6

Scientific Reasoning
Inside, Outside, On, and Beyond the Box

Expected Outcome
Learners are asked to predict the number of beads inside a projected image of an opaque box (or an actual 3-D box) based on an observable pattern of beads on a string that appears to go in one side and out the opposite side of the box.

Science Concepts

Direct observation, measurement, indirect evidence, and inferential reasoning are used to uncover nature's patterns and their underlying logic. Scientists use both data interpolation and extrapolation, as well as new technologies, to find ways to peer into and open nature's structures of black boxes within black boxes. We use similar processes for refining and improving complex human-engineered systems. For example, black box flight recorders on airplanes are designed to help analysts figure out what went wrong ("to probe inside the box") after a malfunction or disaster to reduce future mistakes. Preventative medical diagnostic procedures and autopsies serve a similar function with human health.

Science Education Concepts

Picture puzzles and physical props can be used as *visual participatory analogies* for "black box" systems (BBS) that model the nature of science (NOS). Additionally, what happens inside a learner's mind as a result of interaction with a particular learning environment can be viewed somewhat as a BBS where assessment probes provide diagnostic and formative feedback (Keeley, Eberle, and Farrin 2005; Keeley, Eberle, and Tugel 2007). Similarly, schools as learning organizations for both students and teachers can be viewed as BBS that require continuous cycles of testing and adjustment of hypotheses about student–subject matter relationships. Similarly, learning-teaching theories and practices are always research works in progress.

This activity also demonstrates the pedagogical principle of the power of *multisensory experiences and multiple contexts in activating attention and catalyzing cognitive processing* that helps novices develop more conditionalized, organized, expert knowledge (NRC 2006). The contrast of using a 2-D image versus an actual 3-D box emphasizes the former and the series of Extension activities models the latter. Activities #19 and #20 in *Brain-Powered Science* also feature this principle of constructivist learning theory. "Two by four" teaching—which limits students' experiences to what's between the two covers of the textbook and four walls of the classroom—produces bored students. Students need to experience and reflect on the wondrous emergent properties that exist at each nested system level of reality, from subatomic particles to the

Scientific Reasoning

universe as a whole (Morowitz 2002). Teachers facilitate this by bringing the outside world into the classroom; taking students outside the confines of the classroom for live and virtual field investigations; and using analogies, physical models, multimedia, and other minds-on instructional strategies to help students conceptually connect the different levels of reality.

Materials

A 2-D image of the beads in a box puzzle (illustration on p. 71) to project or use as a handout and/or use an opaque box with two colors of plastic-coated paper clips (or beads) linked in a pattern where colors alternate, with one color increasing incrementally by one extra paper clip with each single paper clip (or bead) of the other color (i.e., 1 B + 1 W + 2 B + 1 W + 3 B on the left side of the box and 6 B + 1 W paper clips (or beads) emerging on the right side of the box). Conceal the portion of the pattern in between that consists of 1 W + 4 B + 1 W + 5 B + 1 W paper clips (or beads) hidden inside the closed box.

> ## Points to Ponder
>
> *If I have ever made any valuable discoveries, it has been owing more to patient attention than any other talent.*
>
> —Isaac Newton, English mathematician, astronomer, and physicist (1642–1727)
>
> *In our endeavor to understand reality we are somewhat like a man trying to understand the mechanism of a closed watch. He sees the face and the moving hands, even hears it ticking, but he has no way of opening the case. If he is ingenious, he may form some picture of a mechanism which could be responsible for all the things he observes, but he may never be quite sure his picture is the only one which could explain his observations. He will never be able to compare his picture with the real mechanism and he cannot even imagine the possibility of the meaning of such a comparison.*
>
> —Albert Einstein, German American physicist (1879–1955)

Procedure

Project the image, distribute a paper handout, or use an actual physical model of the Beads in a Box puzzle (as illustrated on p. 71). (*When working with teachers*: If time permits, you may wish to use each method with one-third of the group for later analysis of their relative advantages.) Ask learners to infer how many beads are in the box and what the implications of this visual participatory analogy are by addressing the following questions:

1. What pattern(s) can you observe by examining the beads that are visible outside the box? If you have a paper copy of the image or an actual box to examine, what additional *empirical evidence* could you gather?

2. Based on your observations, predict how many beads you think are inside the box, counting those that might be hidden from view. Make a drawing of what you think the inside of the box looks like. Propose and draw at least one alternative plausible hypothesis.

3. Explain your reasoning and the assumptions (i.e., *logical arguments*) that underlie your prediction.

4. Can we be sure that the pattern we infer about the inside of the box is real and not an artifact of our creative imaginations? Why or why not (i.e., *skeptical review*)? Would having an actual sealed box that we could manipulate and subject to certain tests make our prediction more certain? What technologies might enable us to peer inside without actually physically opening the box?

5. How is this system an *analogy* for scientific reasoning and the NOS? How does science balance creative imagination and critical review in the generation and testing of hypotheses?

Debriefing

When Working With Teachers and Students

Discuss their reactions to the questions in light of the two quotes on page 73. Identify some real black box systems that will be studied later in the year (e.g., atomic models, the "string of beads" of RNA

74

NATIONAL SCIENCE TEACHERS ASSOCIATION

Scientific Reasoning

and DNA nucleotide sequences, the fossil and molecular records, plate tectonics, climate and weather models, or electrical circuits).

When Working With Teachers

Discuss the differences between using a projected image, a paper copy of the image, or a 3D model. For example, a paper copy allows one to measure the size of the balls and the walls and base predictions on linear measurements, whereas learners with the actual box could also use mass determinations to estimate the number of hidden balls to confirm the color pattern and linear measurements. In any case, consider the idea of a black box system as an analogy for reforming school practices to be more research-informed by discussing open-ended questions such as the following: How can teachers, departments, and schools learn to both probe inside and think outside the box of our current curriculum-instruction-assessment practices and models? How does our framing of pedagogical problems restrict and constrain the range of our solutions? What opportunities for learning and growth are hidden in our pedagogical problems? What special advantages does science have with respect to providing students with multisensory experiences (e.g., hands-on materials, physical models, simulation games, computer animations, and time-lapse and microscopic photography)? How are multiple contexts for learning built into the 5E Teaching Cycle (Engage, Explore, Explain, Elaborate, and Evaluate)? Also, challenge teachers to complete one or more of the Extensions as an example of how multiple contexts can be used to increase the odds that learners will understand and retain a given science concept or process skill and be able to transfer and apply their learning to new but related contexts.

Extensions

The following activities all play off of two analogies—the ideas that the process of science will be related to forever probing black boxes and that our thinking will be confined to certain perceptual and conceptual "boxes" that we need to reach beyond:

1. *Thinking <u>Inside</u> the Box Version #2:* Science supply companies sell versions of the 1960s Elementary Science Study/ESS black

box experiment, in which sealed containers are manipulated by students to form inferences about the inside design and contents of the box. For example, see Lab-Aids Incorporated: #100 *Ob-Scertainer: A Better Black Box*: $79.00 for a set of 24 "boxes" with 12 different internal arrangements (*www.lab-aids.com/catalog.php?item=100*). Alternatively, learners can design and construct their own black boxes and challenge each other to infer the contents without opening them up for direct visual inspection. See Activity #7 in *Brain-Powered Science* and the Internet Connections here (e.g., Mystery [Mailing] Tube) for other examples.

2. *Thinking on the Box*: The National Academies Press book *Teaching About Evolution and the Nature of Science* (1998), Activity #1: "Introducing Inquiry and the Nature of Science" (pp. 66–73) describes a 5E Teaching Cycle activity that uses a constructed paper cube as a mystery box. Opposite sides of the cube contain either a male or female name with a certain number of letters (X) in common. Each side of the cube also contains two numbers in opposite corners; the lower left-hand corner contains the number (X) and the upper right-hand corner contains the number (Y) of letters in the name on that side of the box. The boxes are distributed with one side covered over, and students are asked to discover the patterns and predict what name and numbers are on the covered side of the box. The template for this easy-to-construct inquiry-oriented box is on page 72 in *Teaching About Evolution and the Nature of Science* (*http://books.nap.edu/openbook.php?record_id=5787*).

3. *Thinking Outside the Box*: The Nine Dots Puzzle asks learners to use four straight lines to connect nine dots arranged in a three-by-three square matrix pattern. This puzzle highlights the following:

- the fact that how we see reality is influenced by our prior conceptions; our "mind's eye" influences the apparently seamless, instantaneous translation of raw empirical data from attention or sensation → reception or perception → conception (see Activity #5 in *Brain-Powered Science*)
- the human tendency to impose artificial constraints on solutions to problems

Scientific Reasoning

- the need to be creative and think outside the box of conventional thought when approaching unique problems

In addition to the classic four-line solution, learners can be challenged to discover a three-line solution (*Hint:* If the dots are big enough and the lines extend far enough beyond the artificially imposed box) or a one-line solution (*Hint:* Use a wide paintbrush or accordion-fold the paper to line up the three dots.). See the image above or go to the Internet Connections. Also consider how the image of a box remains even if only four corner dots are shown or if only four L-shaped corners are drawn oriented inward or four lines are drawn in a box shape without the four corner L-pieces; in all cases, most humans still see a box. Even though arguably no box is actually present, we infer the existence of a box.

4. *Thinking of Boxes Within Boxes*: Activity #17 explores the idea of powers-of-ten scales and metric volume measurements with a visual model of nested boxes (mm^3, cm^3, dm^3, and m^3).

5. *Thinking All Around the Box*: The Rubik's Cube is a three-by-three-by-three manipulative puzzle that challenges players to make each of the six sides a different solid color. It can be used to teach a variety of science and mathematics concepts, as well as critical-thinking and problem-solving behaviors (see Internet Connections).

6. *Thinking <u>Within and Across Interconnected</u> Boxes: Crossword Puzzles as an Analogy for Science*: Challenge teachers to (a) complete a simple crossword puzzle that contains terms such as *science, inquiry, data, observation, hypothesis, theory, falsification*, and so on, and (b) consider how a crossword puzzle is an analogy for the nature of science, where productive theories have applications to multiple phenomena and synergistic relationships with other theories (i.e., much as individual letters gain credibility as they are shared by two interconnected words on the crossword puzzle). A particularly compelling case is the contributions of multiple scientific disciplines to our understanding of biological evolution. A variety of free and commercial puzzle-making software can be used to generate puzzle templates (see Discovery Channel in Internet Connections). See also Activity #9, Extension #1 for a jigsaw puzzle analogy. Concept mapping is another tool to visually represent the interconnected web of relationships among concepts as based on constructivist learning theory. See Internet Connections.

Internet Connections

- Biology Corner: Scientific Method (variety of lesson plans and resources): *www.biologycorner.com/lesson-plans/scientific-method*
- Constructivism and Learning Theories: Supplemental readings for teachers: *http://carbon.ucdenver.edu/~mryder/itc/constructivism.html*

    Concept to Classroom: *www.thirteen.org/edonline/concept2class/constructivism/index_sub5.html*

    Encyclopedia of Educational Technology: *http://eet.sdsu.edu/eetwiki/index.php/Main_Page*

    Exploratorium: *www.exploratorium.edu/IFI/resources/constructivistlearning.html*

    Human Intelligence: New and Emerging Theories of Intelligence: *www.indiana.edu/~intell/emerging.shtml#intro*

    Learning-Theories.com: *www.learning-theories.com*

    Wikipedia: *http://en.wikipedia.org/wiki/Constructivism* (learning theory, teaching methods)

Scientific Reasoning

- Discovery (Channel) Education: PuzzleMaker: criss-cross, word search, and other templates: *http://puzzlemaker.discoveryeducation.com*
- Disney Educational Productions: *Bill Nye the Science Guy: Patterns* ($29.99/26 min. DVD): *http://dep.disney.go.com*
- Institute for Human and Machine Cognition: Concept Mapping Tools (free software): *http://cmap.ihmc.us* (Dr. Joseph Novak, the originator of concept mapping, is affiliated.)
- MakeBeliefsComix! (a comic strip generator; select characters and facial expressions and then add your own words in the boxes): *www.makebeliefscomix.com*
- MIT Media Lab's Beyond Black Box (BBB) Project: *http://llk.media.mit.edu/papers/bbb*

    This *Journal of Learning Sciences* (2000) article focuses on tiny, fully programmable devices called Crickets that students can embed in (and connect to) everyday objects to control motors and lights, receive information from sensors, and communicate via infrared light.

- Mystery (Mailing) Tube: *http://undsci.berkeley.edu/lessons/mystery_tubes.html*

    www.bsu.edu/web/fseec/pie/Lessons2/General Science/Mystery Tube.doc and *www.teacherlink.org/content/science/class_examples/Bflypages/timlinepages/nosactivities.htm* (see also Fossil Footprints and Dinosaur Bones activities)

- NASA: A Black Box for People: portable, noninvasive black box to monitor human health: *http://science.nasa.gov/headlines/y2004/07apr_blackbox.htm*
- National Center for Case Study Teaching in Science: Thinking Inside the Box: *http://ublib.buffalo.edu/libraries/projects/cases/box/box1.html*
- Nine Dots Puzzle: visual template and discussion of its significance:

    Brainstorming.co.uk: *www.brainstorming.co.uk/puzzles/lateralthinkingpuzzles.html*

    Wikipedia: *http://en.wikipedia.org/wiki/Outside_the_box*

- *Research Matters—to the Science Teacher*: *http://narst.org/publications/research.cfm*

See, for example, constructivism and the learning cycle, conceptual change teaching, metacognitive strategies, and pedagogical content knowledge.

- Rubik's Cube:

    Rubik's Official Online Site: *www.rubiks.com*

    You Can Do the Rubik's Cube: *www.youcandothecube.com*

    Wikipedia: *http://en.wikipedia.org/wiki/Rubik%27s_Cube*

- University of Virginia Physics Department: Indirect Measurements and Black Box Experiment: *http://galileo.phys.virginia.edu/outreach/8thGradeSOL/IndirectMeasure1Frm.htm*

Answers to Questions in Procedure, steps #1–#5

1. *Empirical evidence*: From the projected image, learners should be able to observe an alternating pattern of black and white beads with the number of black beads increasing by one with each successive white bead on the two sides of the box. If a paper handout is used, learners could measure one edge of the cubic box and calculate how many standard-size beads would fit along the sides and bottom of the box, assuming the string were draped in that fashion. If an actual box and colored beads or paper clips are used, the string or linked paper clips could be pulled at both ends to check the assumption that it is continuous; the box could be shaken to confirm that beads (or paper clips) are inside; and the mass of an empty box and single bead (or paper clip) could be compared to the sealed box to estimate the number of hidden objects.

2. *Logical argument*: 4 + 5 black beads hidden from sight + the 2 visible white beads + 1 white bead hidden from sight = 12 beads that are likely to be inside the box *if* the pattern holds.

3. *Logical argument* examined through the lens of *skeptical review* requires consideration of assumptions. Assumptions include that there are more beads inside the box beyond the two white ones that are visible; the string of beads is continuous; and the pattern that is visible on the two exterior sides of the box continues on the inside of the box.

Scientific Reasoning

4. Extended *skeptical review* would suggest that based on the projected image alone, we could not be certain the pattern continues. Having a paper copy or 3D box to examine would provide additional opportunities that might either falsify or provide additional confirmatory data. For example, a paper copy would allow measurements of the diameter and spacing of the ball and the cube's dimensions to see how many balls might fit along two walls and the base. Or an actual box could be shaken, weighed, and compared to the mass of one ball to determine how many balls of a given uniform mass might be inside. However, without a means to actually open the box and peer inside, we could never be 100% certain our prediction was correct.

5. Science is the search for and explanation of patterns in nature. Scientists often work with opaque, sealed black boxes that they can't directly peer into or open with current technologies. In such cases, they make inferences based on other empirical evidence and hold onto their theories or models as provisional truth until further experiments prove them wrong or a more elegant theory is developed that better explains the available data or makes more useful predictions for additional experimentation. Creativity has been described as thinking outside of the box (see Extension #3), but it can also involve making interpolations about the inside of the box. Science is a process that sometimes combines a wild and crazy, creative imagination that is bounded by critical experimental testing against the reality of nature and collective, skeptical review of the results by an international community of scientists.

Magic Bus of School Science
"Seeing" What Can't Be Seen

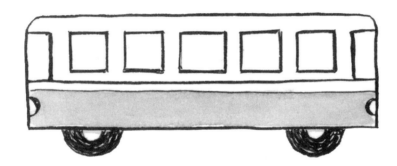

Expected Outcome

A simple line-sketch drawing of a school bus cues learners to infer the presence of an entry or exit door on the opposite side of the bus (not visible in the drawing) to predict the direction in which the bus is moving.

Science Concepts

Scientists use direct empirical and inferential evidence, logical argument, and skeptical review to formulate and evaluate hypotheses and tentative, provisionally accepted theories. The questions asked, observations made, and conclusions drawn are all influenced by prior understandings of individual scientists and the broader scientific community. Assumptions, a necessary part of the nature of science (NOS), need to be acknowledged and critically reviewed, especially when exploring "black box" systems at the limit of our ability to make direct observations. Extensions on pages 88–89 suggest investigations related to directed movement (or taxis) in mealworms and STS issues related to mass transportation.

Science Education Concepts

A real school bus gets students to the science classroom; the direction that they go from there depends on the intelligence of the curriculum-instruction-assessment (CIA) system in place. The *Benchmarks for Science Literacy* (AAAS 1993), *National Science Education Standard/NSES* (NRC 1996), and *A Framework for Science Education, Preliminary Public Draft* (NRC 2010) call not for simply increasing the speed of the bus for a change in direction to improve school science. This activity serves as both a *visual participatory analogy* for and a concrete example of the desired change in direction toward more inquiry-oriented, minds-on, thought-provoking CIA. No cost, minimal preparation paper-and-pencil puzzles can be used on a regular basis as a complement to more involved hands-on explorations—both challenge learners to "hoe the garden" of their minds in preparation to support the seeds of new ideas. Activities #10–#17 in Section III examine the idea of curriculum standards or desired directions in more depth.

This discrepant-event activity also demonstrates the pedagogical principle of the power of *emotional engagement, connections,* and *relevance* in *activating attention and catalyzing cognitive processing* (Gardner 1999; Goleman 1995; McCombs and Whisler 1997). Activities #21 and #22 in *Brain-Powered Science* also feature this principle of constructivist learning theory. School buses evoke memories and feelings that help set a relevant, everyday context for learners to consider

Magic Bus of School Science

the more abstract NOS as a way of knowing. The picture puzzle also helps introduce the idea of science as fun work and hard play where humor and laughter (*ha-has*) can lead to critical insights (*ahas*).

Materials

- Overhead transparency, PowerPoint slide, or handout drawing of a crude, symmetrically shaped sketch outline of the lengthwise side view of a school bus. The bus should lack a clearly distinguishable front and back (i.e., neither the driver nor entry door should be visible), but it should have visible windows and two tires (e.g., a simple, symmetrical, hot-dog-shaped body with several windows and two wheels). This activity is modified from the Preschooler's Test that has been circulated anonymously over the internet; see, for instance, *www.dwlz2.com/forum/showthread.php?t=2046* and *www.drunken-pumpkin.com/2006/10/visual-puzzle-a-school-bus*
- *Optional Variations:* Music: Theme song from the *Magic School Bus* series.

Also, Activity #12's combination of "disappearing" phenolphthalein ink and household ammonia developing spray can be used to temporarily reveal an answer arrow that points in the correct direction on a white poster that contains the image of the bus.

Procedure

1. Project an image or display a poster of the simple symmetrical school bus (with the accompanying song from the *Magic School Bus* series if desired) and use a think-write-pair-share to ask learners to first individually and then collectively (in dyads or triads) address the following questions:
   a. Based on the available *empirical evidence* (observe), in which direction (predict to the left or right) can you infer that the bus is headed?
   b. What *logical arguments* (explain) support your inference?
   c. What assumptions have you made (i.e., need for *skeptical review*)?

Points to Ponder

Now, here, you see it, it takes all the running you can do, to keep in the same place. If you want to get somewhere else, you must run at least twice as fast as that.

—The Red Queen in *Through the Looking Glass* by Lewis Carroll, British mathematician and author (1832–1898)

In gaining knowledge you must accustom yourself to the strictest sequence. You must be familiar with the very groundwork of science before your try to climb the heights. Never start on the "next" before you have mastered the "previous."

—Ivan Pavlov, Russian physiologist (1849–1936)

Absence of evidence is not evidence of absence.

—Sir Martin Rees (1942–), British astrophysicist, cosmologist, and president of the Royal Society (December 2005–)

   d. What role does prior experience with school buses play in your ability to solve this puzzle? If necessary, provide a hint by way of the quote from Sir Martin Rees; empirical data may include both what we see and don't see, and logical argument may include attention to what is missing but needs to be there.

2. When Working With Teachers

   a. How does the *National Science Education Standards* emphasis on the NOS as defined by *empirical data and evidence, logical argument,* and *skeptical review* and unifying concepts and processes (NRC 1996, pp. 115–119) help set the direction for school science education? What current practices might we have to stop and reverse direction on (instead of running twice as fast in the same direction) to better align local curriculum-instruction-

Magic Bus of School Science

assessment with the *NSES* and ensure that all students can board the bus and make the journey to its intended destination?

b. The AAAS *Benchmarks* and NRC *National Science Education Standards* (see Internet Connections) outline grade-level-linked developmental sequences (K–2, 3–5, 6–8, and 9–12 and K–4, 5–8, and 9–12, respectively) for a deepening, broadening, and evolving understanding of core science concepts. Research on and development of learning progressions are ongoing. How does the idea of an articulated, intra- and intergrade-level curriculum scope and sequence relate to Pavlov's quote?

Debriefing

When Working With Teachers

Beyond the discussion questions provided in the Procedure section, make note of the importance of prior knowledge and emotional engagement, connections, and relevance in problem solving and constructing new insights. Building a solid experiential and conceptual foundation and scaffolding is one reason why intra- and intergrade/K–12 scope and sequencing are so important in teaching science for understanding. Professional organizations and researchers are increasingly focusing on the idea of research-validated learning progressions, especially when it comes to teaching big idea theories such as atoms, cells and evolution, plate tectonics, and energy conservations and conversions, which were initially considered quite outrageous concepts.

Teachers may want to read books and articles that feature the unusual nature of science (see Activity #2, Extension #1). Also, the Internet Connections contain several websites that offer assessments that measure understanding of the nature of science.

When Working With Students

Discuss how the nature of *empirical evidence, logical argument,* and *skeptical review* will be a continuing theme in the science course, and how sometimes what's missing or not observed can offer important clues to help solve a scientific mystery. Learning science involves combining FUN (play) and MENTAL (work) on a daily basis. *Optional Discrepant Demonstration*: If a poster (rather than a projected) image

of the bus that has been pretreated with phenolphthalein ink drawn in the shape of an arrow is used, the correct answer can be temporarily displayed (it will fade back to colorless) if the poster is sprayed with household ammonia. (See Activity #12, "Magical Signs of Science," for directions and safety precautions.)

Extensions

1. *TV That Teaches*: Science education PBS television and video series such as the *Magic School Bus* (elementary); *Bill Nye the Science Guy* (grades 3–8) and *The Eyes of Nye* (grades 7–12); and *NOVA, NOVA NOW, Scientific American Frontiers*, and *Wired Science* (high school) commonly feature the nature of science in creative, engaging ways. Also, for high school classes, carefully selected video clips from commercial TV shows such as *Bones, CSI, Numbers*, and *Lie to Me*, or programs on cable channels such as the Discovery Channel (e.g., *MythBusters*) and National Geographic Society can be used to challenge students to separate science fact from fiction and to highlight how science and mathematics help solve real-world problems. See Internet Connections.

2. *Mealworm Movements Matter*: Mealworms are a convenient organism to use to study the biological phenomenon of taxis, the directional orientation and movement toward desired or away from undesired stimuli. See Internet Connections.

3. *STS Issues Related to Moving People*: Investigate the tradeoffs of (a) natural gas or hybrid versus conventional diesel, school, and city buses; and (b) high-tech public transportation versus overreliance on cars and buses. Research and debate the relative benefits and costs of alternative directions for re-engineering public transportation in the United States.

4. *Inattentional Blindness: Not Seeing What Can Be Seen*: The ability to intentionally focus our vision and mental attention means that parts of our environment fall outside the scope of our "attentional spotlight" by default. In magic tricks and everyday life contexts, we can be "blind" to phenomena right before our eyes, even when they change if we are selectively focused on something else (Chabris and Simons 2010; Macknik and Martinez-Conde 2010). For instance, when asked to look at a video and

count the number of basketball passes (or ball bounces) made by one team, most people fail to see a man in a gorilla suit walk in and out of the scene (e.g., *www.theinvisiblegorilla.com/videos.html*, or search YouTube videos for "inattentional or change blindness" or "selective attention tests").

Internet Connections

- American Association for the Advancement of Science, Project 2061. 1993. *Benchmarks for science literacy*. New York: Oxford University Press. *http://project2061.aaas.org*.

- *American Educator*, journal of the American Federation of Teachers: Spring 2009 issue: *Why don't students like school? Because the mind is not designed for thinking* (book): *www.aft.org/pubs-reports/ american_educator/issues/spring2009/index.htm*

- Bill Nye, the Science Guy, Nye Labs: *www.billnye.com*

    http://dep.disney.go.com/educational/search?form. keywords=Bill+Nye (100 episodes)

- *MythBuster*. *http://dsc.discovery.com/fansites/mythbusters/about/ about.html*

- Mealworm Experiments (i.e., Taxis or movement toward or away from external stimuli):

    www.smithlifescience.com/MLMealworms.htm (with links to other sites)

    http://caspar.bgsu.edu/~courses/Ethology/Labs/Taxis

- National Research Council (NRC). 1996. *National science education standards*. Washington, DC: National Academies Press. *www.nap. edu/openbook.php?record_id=4962*

- National Science Teachers Association (NSTA). 2000. *The nature of science*. An NSTA Position Paper. Arlington, VA: NSTA. *www. nsta.org/about/positions/natureofscience.aspx*.

- Nature of Science (NOS): Student Assessment Instruments:

    Draw-A-Scientist Test: *www.theaps.org/education/2006rts/pdf/ DASTRatingRubric.pdf*. See also: *http://findarticles.com/p/ articles/mi_qa3667/is_200211/ai_n9160846*

Nature Of Science Literacy Test: www.phy.ilstu.edu/jpteo/NOSLit.pdf

35 multiple-choice and true/false items; copyrighted but contact *wenning@phy.ilstu.edu* for password and permission to use. See also related articles in the *Journal of Physics Teacher Education Online* by Carl J. Wenning: (a) "A Framework for Teaching the Nature of Science," 3 (3): 3–10. www.phy.ilstu.edu/jpteo/issues/mar2006.html; and (b) "Assessing Nature-of-Science Literacy as One Component of Scientific Literacy," 3 (4): 3–14. www.phy.ilstu.edu/jpteo/issues/sum2006.html

Relevance of Science Education (ROSE): International look at affective factors in learning: www.ils.uio.no/english/rose/about/rose-brief.html

Views on the Nature of Science (VNOS) Questionnaire (open-ended essay questions): *www.eric.ed.gov/ERICWebPortal/custom/portlets/recordDetails/detailmini.jsp?_nfpb=true&_&ERICExtSearch_SearchValue_0=ED472901&ERICExtSearch_SearchType_0=no&accno=ED472901*

Views on Science and Education (VOSE) Questionnaire: *www.ied.edu.hk/apfslt/v7_issue2/chensf/index.htm*. VOSE consists of 15 questions, each followed by between three and nine items representing different philosophical positions. Participants rank each item on a five-point Likert scale. The original source is Chen, S. 2006. "Development of an Instrument to Assess Views on Nature of Science and Attitudes Toward Teaching Science." *Science Education* 90 (5): 803–819.

- View on Science-Technology-Society (VOSTS; 113 multiple-choice questions):

    www.usask.ca/education/people/aikenhead/vosts.pdf

    www.usask.ca/education/people/aikenhead/vosts_2.pdf

- NOVA: *www.pbs.org/wgbh/nova*. Many titles feature the NOS including: *Anastasia: Dead or Alive?* and *Secrets of the Psychics*.

- Scholastic's *The Magic School Bus: www.scholastic.com/magicschoolbus*

    Made-for-TV and chapter books, TV shows, videos and DVDs, games, and experiments that feature inferential

Magic Bus of School Science

reasoning. Although targeted for elementary grades, the materials are also useful for middle school.

- *Scientific American Frontiers: www.pbs.org/saf.* Features the nature of science; see, for example, the show *Beyond Science* (pseudoscientific claims).
- Wikipedia: *http://en.wikipedia.org/wiki/Main_Page*. Search topics: change or inattentional blindness, history and philosophy of science

Answers to Questions in Procedure, steps #1 and #2

1. Assuming that the bus is going forward rather than in reverse, the absence of a visible side entry or exit door implies that the bus is headed to the left because the entry or exit door is always on the right side of the bus from the perspective of the driver and passenger seats. Also, in the United States, we drive on the right side of the road with the driver's seat farthest away from the curb and in the front of the bus, and students are picked up at the curb rather than in the middle of the street. It is likely that learners who regularly wait in a line to ride buses would think about the absence of a visible entry or exit door as an important bit of missing information.

2. *When Working With Teachers*: The saying "less is more" relates to the NRC's *NSES* and the AAAS *Benchmarks* recommendation that we cover fewer topics in more depth to avoid the "mile wide, inch deep" curricula that are most common in schools today. Selection of the "big picture forest" of the most central, powerful explanatory theories, principles, unifying themes, and overarching concepts helps teachers avoid trying to quickly cover every "tree" or factoid in their textbooks. Equally important is the research-based sequencing of intra- and intergrade level learning progressions to develop both a deepening and broadening conceptual understanding of science that makes sense to students. The idea of a spiral curriculum has philosophical roots that predate Pavlov by centuries. Understanding learning as a construction process offers a visual analogy of assessing and turning over the ground of prior conceptions and then building a firm foundation rather than layering excessive, unsupported

weight on top of a shaky base. Helping learners recover valid conceptual precursors (to build upon), uncover misconceptions (to correct), and discover powerful scientific ideas and "ways of knowing" are more important than merely "covering" content. Curriculum-embedded diagnostic, formative, and summative assessments (see Activities #21 and #22) provide feedback to both teachers and their students as to when to adjust the pace and how to vary the curriculum and instruction. These research-informed changes are necessary to achieve the goal of science for all Americans.

Reading Between the Lines of the Daily Newspaper

Molecular Magic

Expected Outcome

Standard-size newspaper has a grain that causes it to tear more readily in the direction perpendicular to the line of printing (vertical) than in the direction parallel to it (horizontal). In fact, horizontal tears will "magically" veer toward the vertical direction.

Science Concepts

Although there is no single scientific method, scientists use the triple tests of empirical evidence, logical argument, and skeptical review to discover patterns in nature and explain their underlying reasons in terms of naturalistic theories that remain open to further testing. Technological aids often are used to extend the limitations of human sensory systems and allow for further testing and refinement of theories. Trees (and paper manufactured from trees) have a natural grain due to the long biomolecular polymers of which they are composed and that run predominantly in a vertical direction.

Evidence for the nanoscale world of atoms and molecules can come from a careful study of even simple physical changes such as tearing newspaper. Many everyday phenomena, including both physical and chemical changes, only make sense when viewed through the lens of the kinetic molecular theory (KMT). The KMT, like all good theories, has both great explanatory and predictive power. The forms (or structures) of both living organisms and manufactured products are designed to match their functions. Given that nature always conserves and recycles atoms, reusing and recycling manufactured products makes sense from an eco-friendly, science-technology-society (STS) perspective (see Extension #2).

Science Education Concepts

Inquiry skills and fundamental concepts (e.g., the atomic and molecular nature of matter) can be presented in the classroom using low-cost everyday materials and phenomena. Nature's patterns are ubiquitous but not always immediately obvious to the untrained or unaided eye. In an analogous manner, students' prior conceptions are like molecular patterns and orientations in that they often lie out of sight and need to be activated and brought into the light of day to be tested. Students need to be taught to predict-observe-explain to read between the lines of their science textbooks and laboratory activities to truly understand (rather than blindly accept, memorize, and mindlessly regurgitate) the principles and nature of science (NOS).

Beyond activating attention and catalyzing cognitive processing in learners, effective curricula provide *adequate time for learning* the big, often counterintuitive ideas of science (AAAS 1993; NRC

Reading Between the Lines of the Daily Newspaper

1996, 2010). Rather than flipping through the pages of the textbook, students need time to read and reflect on nature from multiple perspectives. Standards-based, research-informed best-practice teaching demands a rational sequence of planned steps both within a given course (e.g., the 5E Teaching Cycle, Bybee et al. 2006) and across grade levels (i.e., learning progressions). As such, this activity serves as a *visual participatory analogy* to discuss the need for vertical, cross-grade-level (K–12) articulation and horizontal, within-course and grade level, cross-discipline integration as promoted by the *National Science Education Standards* (NRC 1996), *A Framework for Science Education, Preliminary Public Draft* (NRC 2010), and *Benchmarks* (AAAS 1993) documents. Reducing the scope of the curriculum by covering fewer topics and terms allows teachers to guide their students more effectively to uncover misconceptions and discover the core concepts and overarching theories in more depth. This focus on helping students "see the forest for the trees" results in deeper conceptual understanding, longer retention, and increased ability to transfer and creatively apply learning to new contexts. *Note:* Activities #23 and #24 in *Brain-Powered Science* also feature the principle of less is more and the idea of providing adequate time for learning.

Materials

- Standard-size, double-width (24 in. wide, 21 in. tall) newspaper
- *Optional*: hand lenses or 30X handheld microscopes, celery (or wood log), and cooked spaghetti
  - Edmund Scientific's: Illuminated 30X Pocket Microscope, #X30350-01 for $10.95 (two AA batteries not included): *www.scientificsonline.com*.
  - *Horticulture Source*: EcoPlus Illuminated Microscope 30X, #704477 for $17.95 (uses 2 AA batteries): *www.horticulturesource.com/product_info.php/products_id/1194*
  - *Radio Shack*: Illuminated 60-100X Microscope Model: MM-100, Catalog #: 63-1313 $12.09; *www.radioshack.com* (*Note:* Higher magnification microscopes have a restricted field of view that amplifies the effect of slight hand movements.)

Safety Note
Food should never be consumed in a laboratory.

Activity 8

Points to Ponder

I would rather discover one scientific fact than become king of Persia.

—Democritus, Greek natural philosopher and early atomist (470–380 BC)

Reading furnishes the mind only with materials for knowledge; it is thinking [that] makes what we read ours.

—John Locke, English philosopher (1632–1704).

To see a World in a grain of sand, / And a Heaven in a wild flower, / Hold Infinity in the palm of your hand, / and Eternity in an hour ...

—William Blake, English poet, painter, and printer (1757–1827), "Auguries of Innocence"

All things in nature, by immortal power, near and far; hiddenly—to each other linked are, that you cannot stir a flower, without troubling a star.

—Francis Thompson, English poet (1859–1907)

Procedure

1. Tear standard-size, full, double-width sheets of newspaper along the vertical fold to form a stack of uniform 12 in. wide by 21 in. tall sheets. Distribute one sheet to each learner. Introduce the activity as one that will focus on observation skills (empirical data), logical argument (e.g., inferential reasoning, hypothesis formation, principles of a fair test, and analogical reasoning), and skeptical review—the triple tests that form the foundation of science.

2. Define the horizontal tear as running parallel to the line of text and a vertical tear as running perpendicular to it. Ask learner dyads to systematically predict-observe-explain the behavior of

Reading Between the Lines of the Daily Newspaper

the newspaper as it is ripped in horizontal versus vertical directions relative to the print. The two half sheets of newspaper should provide the dyads plenty of material to test. Consider the following questions:

   a. Will the direction of the tear make a difference, and if so, how and why?
   b. What experimental variables should be tested or controlled (held constant) for any tests to be considered fair?

3. Ask dyads to develop a list of observations to compare and contrast the horizontal versus the vertical tears. Challenge them to be as specific as possible in their language and to use as many senses as possible. If available, distribute hand lenses or handheld microscopes after they have made their first rounds of observations. Poll and visually display the class results, refining the descriptive language based on input from all the groups.

4. After the observations are tallied and discussed, make the point that scientists always try to probe the underlying explanatory causes (i.e., *logical arguments* based on unifying theories) for the observed regularities and pattern in nature. Consider such questions as the following:

   a. Are these patterns due to pre-existing natural causes or are they an artifact of the way paper is manufactured?
   b. What is paper made from and how is it made?
   c. Do trees have a natural vertical orientation or grain and if so, why? Consider the different force needed to split logs vertically versus chopping them horizontally.
   d. If wood naturally has a grain, how does the manufacturing process for making newsprint paper preserve this grain?
   e. Why would all full-size newspapers be printed in a way that the vertical direction (rather than the horizontal one) is the easy tear?

Optional Teacher Demonstrations: Celery stalks can be sliced in the vertical versus cut in the horizontal direction to simulate the relative force needed to split or slice versus chop wood logs. In both cases, slicing with the grain is easier than chopping across the grain or fibers (even though logs are typically longer than they are wide). Also, the manufacturing process of turning paper pulp into newspaper can be simulated with a macroscopic, analogy-type demonstration: Slide a mound of wet spaghetti (that models the long cellulose fibers jumbled in a paper pulp slurry) between your hands, moving them in opposite directions against each other, and the spaghetti will tend to align in the direction of movement.

5. *Skeptical review*: Invite learners to use books and internet sources (or contact a local newspaper printing facility) to check out the veracity of the previously discussed scientific explanations. Also, test the half-size Sunday insert news magazine or tabloids to see how those sheets tear. Should the paper tearing patterns be the same in half-size newsprint? Why or why not? Would there be any possible advantage in reversing the pattern for half-size newsprint? If desired, students can also explore the different designs of saws that are used to cut wood with the grain (rip saws) versus across the grain (cross-cut saws) as another example of a form to function fit.

Debriefing

When Working With Teachers

Discuss how this activity is a model for simple-to-do, no-cost but conceptually rich, inquiry-based hands-on explorations. Doing real science does not always require being scheduled for the laboratory room. Quickly cover the science underlying the activity (as need demands and time permits) to reserve time to discuss the science education concepts and quotes listed above. Consider the scientific wisdom captured in the quotes by the poets Blake and Thompson (i.e., atoms are much smaller than "a grain of sand," originated in stellar nuclear furnaces, and holographically contain much of the truth of nature in their structure). This activity also serves as a *visual participatory analogy* to challenge teachers to consider the following concepts:

a. Students need time to make and process observations (including instrument-assisted ones) from multiple encounters with a given phenomenon in order to expand their prior experiential foundations and challenge their preexisting conceptual frameworks. Expecting students to truly understand (rather than blindly

Reading Between the Lines of the Daily Newspaper

accept) the big ideas in science (e.g., kinetic molecular theory) based on quick, one-shot curricular coverage (i.e., tearing or ripping through the pages of the textbook) is the equivalent of asking them to be intellectual supermen or superwomen who can leap tall buildings in a single bound. Core theories in science typically took hundreds of years to develop, with multiple researchers using iterative cycles of empirical evidence, logical argument, and skeptical review to evolve to their current forms. Students deserve the opportunity to develop a sense of how we know what we know, think and act like scientists, and experience the joy of new (for them) discoveries.

b. The slow historical evolution of scientific ideas and the related research on students' often unseen but tenacious misconceptions (see Appendix A) point to the need for carefully designed *learning progressions* that cut across in both the *horizontal* direction (integrated and iterative units within a given course across an academic year) and *vertical* direction (*articulation* across K–12 grade levels) if students are to learn the big ideas in science in a way that does justice to the nature of science and learning. Curricula should not be overstuffed and malnourished as a result of being too busy naming and cataloging all the "trees" so that students miss the "forest." Learning that lasts takes time, and effective teaching requires intelligent, integrated curriculum-instruction-assessment plans that aim to develop higher-level thinking skills (see Internet Connections: Bloom's taxonomy and *Changing Minds* Questioning Techniques). Especially when teachers ask challenging questions, students need wait time or "pause to ponder" time. See Internet Connections: P16 Science Education.

When Working With Students

The tripod foundation of science should be explicitly discussed with students. From a content perspective, this activity is suitable for almost any phase of the 5E Cycle. If it is used as an Engage-phase activity, Activity #15 can be used as the Explore phase leading into the Explain phase, where the idea of the particulate, molecular view of matter is formally introduced. Alternatively, with much less guiding, students should be able to explain the underlying reason for the newspaper pattern as a real-world application during the Elaboration phase of a 5E unit on molecules. (See also Activity #20 in *Brain-Powered Science* as an additional Engage-phase activity). Research on student misconceptions related to foundational chemistry concepts can be found in Driver et al. (1994; chapters 8–11) and Kind (2004).

Extensions

1. *Colorful Comics: Micro ← → Macro Worlds Interactions:* Students can explore the microscopic reality that exists just below naked-eye vision that creates our macroscopic sense of color by using 30X handheld microscope (or even an 8X magnifying lens) to look at colored comic strips from the Sunday edition of their local newspapers. The colors that we see with our naked eye in comic strips and many other examples of color printing are actually made up of an array of different-color, closely packed dots whose colors blend when viewed without magnification and enhanced resolution. Higher-quality color printing has more densely packed, smaller points of color. Students can use the internet to find out more about color printing.

    This playful experience provides another example of how a unique microstructure underlies our macroscopic sense of reality and how things that appear to be solid can in fact be made up of small, unseen particles separated by empty space (i.e., the molecular perspective). Note the actual nanoscale (10^{-9} meters) world of atoms and molecules is orders of magnitude smaller than what students can see with a simple hand lens or microscope. In fact, atoms cannot be seen with even the most powerful light microscope because individual atoms are smaller than the wavelengths of light and therefore cannot reflect light. Electron microscopes allow us to "see" atoms, although believing or interpreting the images requires the viewer to accept the underlying theory of atoms with subatomic particles (including electrons themselves)! Science and technology have greatly extended the range of human senses and altered the meaning of the word *observations*.

2. *Old News Into New: A Science-Technology-Society (STS) Recycling Lesson:* Given the amount of torn-up newspaper produced and the theoretical explanation for the tearing patterns in the main

Reading Between the Lines of the Daily Newspaper

activity, it may be timely to introduce the idea of the law of conservation of matter and recycling. Paper is both a renewable (i.e., new trees can be continually grown and harvested) and a recyclable natural resource. Given that paper is renewable, some students may question the need for recycling instead of simply throwing it away (but "not in my backyard"). Before you discuss the wisdom of recycling, students should have direct experience with the process.

Old, discarded newspaper can be crudely recycled by cutting up pieces, soaking them in warm water, pulping in a blender, adding some bleach to whiten, pressing against a strainer or a screen, rinsing out the bleach, and leaving the pulp to dry. (See Internet Connections: Museum of Science and Industry and Science NetLinks for step-by-step plans for doing this.) Students can explore the quality of such homemade paper and research how paper is recycled on a community-wide scale. A variety of questions can be explored: How many trees does it take to produce a ton of newsprint? How many tons of newsprint are produced in a daily run of a local newspaper? How does recycling paper reduce the need for landfill space? How successful are efforts to encourage recycling in your community? What are the relative pollution and energy effects of recycling versus processing virgin paper from trees? Why is the mantra "Reduce, Reuse, Recycle" in the sequence that it is? How can electronic news venues reduce or replace the use of newspapers? If atoms are always conserved, why is paper quality somewhat degraded with each successive recycling, and how do recycling operations take this into consideration? Also, your students can compare easy-to-tear newspaper to nonrenewable, nonrecyclable (at the present time), tear-resistant DuPont Tyvek. The latter consists of a randomized layout of long, cross-linked polyethylene fibers that are melted together with heat and pressure. Once again, students can consider the applicability of the idea of a form-to-function "fitness" and the idea of a cost-benefit ratio when choosing materials for a given application.

Teachers may wish to read the NSTA and National Association of Biology Teachers position statements on environmental education, sustainability, and STS and discuss with other science teachers ways to incorporate more real-world applications of science into their curricula (see also Section 5 in this book and the Internet Connections).

3. *Water Soluble Paper: An Alternative STS Pollution Solution?* Several suppliers (see Internet Connections) sell a form of paper that is made from sodium carboxyl methyl cellulose and wood pulp that dissolves in water and is advertised as nontoxic, environmentally friendly, and 100% biodegradable. This product can be introduced in a "magic or science?" manner and subsequently used to explore ideas such as form-function fitness (for different applications) and the relative energy, environmental, and economic costs and benefits of direct recycling and reuse (versus allowing nature to do it for us).

4. *Möbius Reads the Newspaper.* The long strips of newsprint that are produced by the vertical tears can be used to explore the phenomenon of single-sided bands. See Activity #2 in *Brain-Powered Science*.

Internet Connections

- American Forest and Paper Association: *www.afandpa.org*
- Bloom's Taxonomy: original and revised/updated summary chart:

    www.kurwongbss.eq.edu.au/thinking/Bloom/blooms.htm

    http://projects.coe.uga.edu/epltt/index.php?title=Bloom%27s_Taxonomy

    www.krummefamily.org/guides/bloom.html

- *Changing Minds* (book and website): Questioning Techniques (19): *http://changingminds.org/techniques/questioning/questioning.htm*

- Chris Jordan, Photographic Arts: Running the Numbers: An American Self-Portrait (interactive art displays the statistics of American life; many relate to environmental issues; see, for example, "Paper Bags," which depicts 1.14 million brown paper supermarket bags, the number used in the United States every hour (*www.chrisjordan.com/gallery/rtn/#paper-bags*); and "Toothpicks," which depicts 100 million toothpicks, equal to the number of trees cut in the United States yearly to make the paper for junk mail (*www.chrisjordan.com/gallery/rtn/#toothpicks*).

Reading Between the Lines of the Daily Newspaper

- Conservatree: Paper making and Recycling: *www.conservatree.com/index.shtml*
- Disney Educational Productions: *Bill Nye the Science Guy*: *Atoms, Garbage,* and *Pollution Solutions* ($29.99/26 min. DVD): *http://dep.disney.go.com*
- Museum of Science and Industry-Chicago: *www.msichicago.org/online-science/activities*

    See How to Make Recycled Paper.
- National Association of Biology Teachers (NABT) Position Statements:

    Sustainability in Life Science Teaching and Teaching about Environmental Issues: *www.nabt.org/websites/institution/index.php?p=35*
- National Science Teachers Association (NSTA) Position Statements:

    Environmental Education: *www.nsta.org/about/positions/environmental.aspx*

    Teaching Science and Technology in the Context of Societal and Personal Issues: *www.nsta.org/about/positions/societalpersonalissues.aspx*
- P16 Science Education at the Akron Global Polymer Academy: Wait Time: *www.agpa.uakron.edu/p16/btp.php?id=wait-time*
- Paper Industry Association Council: *www.paperrecycles.org*
- Ripping into Science: *http://findarticles.com/p/articles/mi_qa3666/is_199505/ai_n8725116*
- Scholastic Recycling Starts with You! (grades 3–6 English/language arts and math lessons): *http://teacher.scholastic.com/lessonplans/recycling*
- Science NetLinks: Paper Making/Recycling (by hand): Grade 3–5 Lesson Plan & Student Handout: *www.sciencenetlinks.com/lessons.cfm?BenchmarkID=8&DocID=27*
- Water Soluble Paper:

    Aquasol Paper: *www.aquasolpaper.com*

Dissolvo Spy paper—Water Soluble Security Paper: *www.kleargear.com/1815.html?gclid=CLjQm_aEzqUCFcbc4AodwW3UlA*

Edmund Scientific's: *www.scientificsonline.com/aquasol-water-soluble-paper.html*

- Wikipedia: Paper, paper recycling, and Tyvek:

    http://en.wikipedia.org/wiki/Paper

    http://en.wikipedia.org/wiki/Paper_recycling

    http://en.wikipedia.org/wiki/Tyvek

    See also: Cross-cut versus rip saw: *http://en.wikipedia.org/wiki/Crosscut_saw*

Answers to Questions in Procedure, steps #2–#5

2.a. Learners probably will not have prior knowledge about the chemical composition of trees and the manufacturing processes used to make paper from trees. However, they may have prior experiences with accidentally tearing newspaper, which may lead them to predict a difference in vertical versus horizontal tears.

b. One control is to always tear the paper in a top to bottom direction. That is, when making a horizontal tear, first rotate the paper 90 degrees. Other possible variables to test: the speed of the tear, its orientation toward or away from the person, how far apart the person's hands are as the tear is made, whether the person's eyes are open or shut (i.e., perhaps the person could unconsciously bias the outcome), the size of paper used (perhaps use presliced 12 in. squares to standardize the lengths of the vertical and horizontal tears), and so on.

3. *Empirical evidence*: Vertical tears are more likely to stay straight and linear and run parallel to each other than are horizontal tears, which curve and, if torn slowly enough, actually veer toward the vertical direction. Relative to horizontal tears, vertical tears appear cleaner, with less layering and hanging paper;

Reading Between the Lines of the Daily Newspaper

sound cleaner (i.e., a narrower range of pitches; an oscilloscope can confirm this); and are easier to make (i.e., take less force and encounter less resistance).

4. *Logical arguments* that account for the behavior of the newspaper when torn include the following:
   a. The tearing patterns are related to both the molecular design of the tree fibers that are the original source of the paper and the manufacturing process that converts paper pulp into newsprint.
   b. Trees consist primarily of long, natural polymers or biomolecules such as cellulose that are oriented largely in a vertical direction to provide strength for the tree to grow upward to maximize access to sunlight for photosynthesis.
   c. Chopping across these fibers takes more force than slicing between them, as when chopping versus splitting logs.
   d. Paper is made from tree fibers that have been broken down into a slurry that is then pressed, dried, and rolled through rollers that cause the long, initially vertical fibers to orient in the direction in which the paper is rolled and dried.
   e. If newspapers were printed so that the fiber orientation was in the horizontal direction, readers would be more likely to rip the paper as they turned or flipped through the pages. As with natural products, human-designed products try to achieve a match between form and function.

5. *Skeptical review*: Given that an individual page of half-size newsprint is nearly square, there is no relative advantage to running the print one way versus the other with respect to ease of tearing as pages are turned. In fact, a half-size page of newsprint is commonly printed in the opposite way than full-size newspaper is printed (i.e., the vertical tear and horizontal tear patterns are reversed). Because there is no functional advantage at the level of the reader-user, the ease and cost of printing must be a deciding factor on how the paper is oriented. Perhaps two pages of half-size newsprint are printed side by side simultaneously. Check this is out with a local newspaper printing facility. The internet can be used to compare and contrast the design of rip versus cross-cut saws.

Pondering Puzzling Patterns and a Parable Poem

Expected Outcome

Various letter or word puzzles (and a jigsaw puzzle and poem in the Extensions) help develop learners' powers of observation, pattern recognition skills, and awareness of perceptual and conceptual biases and the nature of science (NOS).

Science Concepts

Science is built on the foundation of empirical observations and evidence, logical argument, and skeptical review and is subject to continual refinement. The boundaries of knowledge are continually being challenged and expanded by a community of scientists who are spread over time and geographic and international boundaries. Science is a collaborative enterprise where the collective "we" goes beyond the limits of the individual "me."

Science Education Concepts

Students and teachers need playful experiences where they can develop the scientific inquiry skills of observation, pattern recognition, and inferential reasoning and prediction, as well as discover some of the perceptual and conceptual biases of humans. Various kinds of puzzles serve as both *visual participatory analogies* for and models of the processes of scientific inquiry, discovery, and theory building. Our love-hate relationship with puzzles also demonstrates the Goldilocks pedagogical principle—that is, curriculum-instruction-assessment should challenge learners with minds-on, discrepant-event activities that create just the right amount of *cognitive disequilibrium* (or "pain") to motivate students to invest energy into resolving the conflict and earning the *psychological rewards* (or "gain") of having accomplished something new and challenging. (*Note:* Learning often results in a natural biochemical "high" from "feel good" endorphins released in the brain.)

Though real-world relevance is always a plus when teaching science, many people choose to invest time and energy into playing with challenging puzzles just for the fun of it (e.g., crossword puzzle lovers and scientists doing basic research). In most cases, learners commonly do a kind of unconscious cost-benefit analysis to determine whether and how much effort to put into learning. External praise and rewards may or may not contribute positively in this analysis (see Internet Connections: *American Educator*'s "Ask a Cognitive Scientist" and Wikipedia). The key is that the learners receive timely and targeted feedback on their successes and "misstakes." Extension #2 features a parable-poem and serves as a powerful analogy for the nature of science as a collective activity of an

Pondering Puzzling Patterns and a Parable Poem

international community of scientists. It also is an example of how a teacher can integrate English/language art skills and literature into a science classroom. (*Note:* Activities #25 and #26 in *Brain-Powered Science* also feature the idea of a cost-benefit calculus that helps predict learners' motivation, efforts, and outcomes.)

Materials

Variety of puzzles (in the Procedures section) to be projected for whole-class viewing. One or more jigsaw puzzles and a poem are used in an Extension activity.

> ## Points to Ponder
>
> *Science is the attempt to make the chaotic diversity of our sense experience correspond to a logically uniform system of thought.*
> —Albert Einstein, German American physicist (1879–1955)
>
> *Discovery consists of looking at the same thing as everyone else and thinking something different.*
> —Albert Szent-Gyorgyi, Hungarian American Nobel biochemist (1893–1986)
>
> *The most erroneous stories are those that we think we know best—and therefore never scrutinize or question.*
> —Stephen Jay Gould, American evolutionary biologist (1941–2002)

Procedure

Project the following exercises in sequential order for all to see.

Exercise #1

Ask the learners to read the following sentence four times silently. Then ask them to go back and count the number of times the letter

F appears. Ask (and keep a class tally): How many learners counted three *F*s, four *F*s, five *F*s, or six *F*s?

FINISHED FILES ARE THE RESULT OF YEARS OF SCIENTIFIC STUDY
COMBINED WITH THE EXPERIENCE OF MANY YEARS.

Exercise #2
Read the following statements quietly to yourself:

Sometimes your "mind's eye" is
Is quicker than your physical eye.
What do you think might be the
the discrepant thing about this FUNomenon?
Do you usually make the
the same "miss-take" twice?

Exercise #3
Finding Patterns and Extrapolating Data: Identify the next letters in the sequence:

a.   A   E   F   H   I   ___   ___   ___

Optional hint: Consider the letters *not* represented in the sequence.

b.   A   F   H   I   J   ___   ___   ___

Optional hint: Consider the letters as sounds and, again, what's missing.

c.   O   T   T   F   F   ___   ___   ___

Optional hint: Consider the connections between letters, numbers, and words.

d.   Complete this sentence by decoding the last five words:

THE ESSENCE OF SCIENCE   (or "HXRVMXV" in code)   IS   (or "RH" in code)

H V Z I X S R M T   U L I   K Z G G V I M H   R M   M Z G F I V

Pondering Puzzling Patterns and a Parable Poem

Exercise #4

Analyzing Anagrams: A Think-Write-Pair-Share Activity: Divide the class in half. Give half of the class the first set of five scrambled words that can be rearranged to spell out names of animals and the other half the second set of five anagrams that spell out the names of plants, telling the groups only that all the anagrams spell out common English nouns. (Do *not* clue them in to the animal or plant groupings.) Request that the learners work the anagrams in sequential order and only move into a dyad (with someone else in their half of the room) after they have completed at least three of the first five anagrams. Without reviewing their results as a class, ask the dyads to split up and give all learners the challenge of unscrambling the sixth anagram: EAP. Tally the results and compare the relative frequencies of the words *ape* and *pea* in the two halves of the class.

Animal Anagrams
LULB   CALEM   NUKKS   SEUMO   BAZER   EAP

Plant Anagrams
NORC   NOONI   MATOOT   PREPPE   TEBE   EAP

Note: Exercise #4 is adapted from Corey, J. R. 1990. #23: Psychological set and the solution of anagrams. *Activities handbook for the teaching of psychology.* Vol. 3. Washington, DC: American Psychological Association.

Debriefing

When Working With Teachers and Students

Exercise #1

The diversity of responses to the seemingly simple task of counting the number of *F*s in a short sentence creates a discrepant event. (*Note:* Six is the correct answer.) Given that all readers know their letters and how to count, the discrepancy calls for an explanation. Discuss possible hypotheses that might explain this variance. For instance, fast readers are not apt to focus on individual letters or small words such as *of*. Also, the word *of* sounds like *ov* and is likely processed and/or stored as a sound in auditory memory. Asking learners to read the sentence several times before asking them to count the number of *F*s has the effect of defining the task or perceptual set as one of reading, and this becomes a boring activity for good readers, who may tune out mentally. As a test of this hypothesis, ask the learners to think of another letter that shows up in small words and has an alternative sound. For example, *s* can sound like the letter *z*. Count the number of times the letter *s* appears in this statement: "Sunday is shown after Wednesday, isn't it, inside those calendars he has made us paste together for our game days next season?" The letter *s* appears 14 times, but most people will count fewer.

Exercise #2

This exercise also points to the need for careful observing and the common tendency to see what we expect to see and overlook details that don't fit our preconceptions. Similar results occur at *http://planetperplex.com/en/item30*, where learners see the familiar Coca Cola logo when in fact the image reads *Coca Coca*, and at *http://planetperplex.com/en/item131*, where they do not see the second *the* in the phrase "A Bird in the the Bush." The website Invisible Gorilla (*www.theinvisiblegorilla.com/videos.html*) and a YouTube search on "change and inattentional blindness" include videos that demonstrate a related phenomenon.

Exercise #3, a–d

Each of these puzzles involves discovering and extrapolating an underlying pattern. The heuristic process of looking for missing data often is helpful. Code-making and code-breaking have been major endeavors for governments and their militaries for centuries. The breaking of the German WWII code (with Enigma) by British and American scientists and mathematicians helped win the war for the Allied forces and spurred the development of computers (see National Science Center, NOVA, and Wikipedia in Internet Connections). Many daily newspapers run a column called "CRYPTOQUOTE" that asks readers to break a code and decipher the hidden message.

Exercise #4

Analyzing Anagrams: This exercise points out the powerful effect of prior knowledge and experience in forming mental blinders that alternately can be viewed as providing a sharpened focus or lens or as a perceptual bias that limits our understanding. Similarly, existing theories in science can both focus and restrict our vision of the possible.

Pondering Puzzling Patterns and a Parable Poem

Extensions

1. *Jigsaw Puzzle Analogy for Science*: The process of science (and specific fields such as paleontology) can be considered analogous to assembling a jigsaw puzzle where what it is supposed to look like is at least somewhat unknown, some pieces are missing, and some pieces may belong to another puzzle! The process can be modeled with students by giving each of six teams a small number of puzzle pieces that may or may not all fit together (if desired, include pieces from a different puzzle as discrepant data); allow the different teams to collaborate and share their "data," but arrange the total data set so that the puzzle is still incomplete. As is true with jigsaw pieces, the separate pieces of data in science (as well as the theories that are derived from them) should fit together and interlock in mutually supportive ways to form a coherent image that tells a story that makes sense. Ask individual teams and collectives to make predictions based on extrapolation and interpolation as to what the complete puzzle should look like. Then allow them to go searching for missing data in a box that also contains pieces from a different puzzle. Either large, simple children's floor puzzles (approximately 50 pieces) or more challenging tabletop puzzles can be used. Individual students can also work with interactive online puzzles at the BBC Science and Nature website (see Internet Connections).

2. *A Parable Poem*: *Six Blind Men and the Elephant*: John Godfrey Saxe (an American lawyer, poet, and newspaper editor who lived from 1816–1887) translated this ancient Indian fable into the form of a poem that succinctly points to the limitations of human sensory perception and the probability of misinterpretations, erroneous conclusions, and partial truths. As such, it is also an apt metaphor for the international community of scientists and the evolution of science over time. The poem and related visual images can be found at various Internet sites (see Elephant and the Blind Men Poem under Internet Connections).

    After reading the poem, discuss the following open-ended questions:

    1. In what ways can individual scientists be considered "blind"?

2. What does this poem suggest about individual cognition, biases, and sensory limitations?
3. How do new technological developments improve scientists' "vision" over time?
4. How do new theories sometimes serve as more useful "glasses" or "lenses" for scientists?
5. Why is collaboration within and across teams of scientists (over geographic space, national boundaries, and historical time) important for the advancement of science as a field of inquiry?
6. How does this poem relate to the Szent-Gyorgyi quote?
7. In what ways are the "artistic creations" of science similar to and different from paintings, sculptures, and other conventional forms of art? Is a given creation in science ever truly complete? Activity #1 in *Even More Brain-Powered Science* (2011) explores the art-science connection in greater depth.

Internet Connections

- *American Educator,* "Ask a Cognitive Scientist" column: *www.danielwillingham.com*

    How Praise Can Motivate—or Stifle: *www.aft.org/pubs-reports/american_educator/issues/winter05-06/cogsci.htm*

    Should Learning Be Its Own Reward?: *www.aft.org/pubs-reports/american_educator/issues/winter07_08/scientist.htm*

    Caution: Praise Can Be Dangerous (download PDF by Carol S. Dweck): *www.aft.org/pubs-reports/american_educator/spring99/index.html*

- BBC Science and Nature: Online jigsaw puzzles:

    Prehistoric Life Skeleton: *www.bbc.co.uk/sn/prehistoric_life/games/skeleton_jigsaw*

    Solar System: *www.bbc.co.uk/science/space/playspace/games/jigsaw/jigsaw.shtml*

- Disney Educational Productions: *Bill Nye the Science Guy: Patterns* ($29.99/26 min. DVD): *http://dep.disney.go.com*

Pondering Puzzling Patterns and a Parable Poem

- Discovery (Channel) Education: Interactive online puzzles and PuzzleMaker:

    http://science.discovery.com/puzzles/puzzles.html

    http://puzzlemaker.discoveryeducation.com (criss-cross, word search, and other templates)

- Elephant and the Blind Men Poem:

    http://hinduism.about.com/od/hinduismforkids/a/blindmen.htm

    www.biologycorner.com/lesson-plans/scientific-method (go to "The Elephant Poem")

    www.noogenesis.com/pineapple/blind_men_elephant.html

    http://en.wikipedia.org/wiki/Blind_men_and_an_elephant (background on the original story)

- Enigma Cipher Machine (history): *www.codesandciphers.org.uk/enigma*

- Enigma Cipher computer simulator (downloadable): *www.xat.nl/enigma*

- National Science Center: Science Soup: Secret Codes and Ciphers: *www.nscdiscovery.org/TeacherTools/topic_details.asp?ID=295*

- NOVA, PBS Show: *Decoding Nazi Secrets*: *www.pbs.org/wgbh/nova/decoding*

- The Earth to Class, Lamont-Doherty Earth Observatory: Lab: "Science Is Like a Puzzle" activity: *www.earth2class.org/docs/tchrlessonplans/sherwood/science_is_like_a_puzzle.php*

- Wikipedia: *http://en.wikipedia.org*. Search: change or inattentional blindness, cryptanalysis or code-breaking, expectancy theory, expectancy value theory, and cognitive biases

Answers to Questions in Procedure

Exercise #1

There are six *F*s in the sentence.

Exercise #2

Repeated words are *is is*, *the the*, and *the the*.

Exercise #3

One might argue that there could be more than one right answer to these puzzles, but simplicity and elegance are valued in science.

a. *B*, *C*, *D*, and *G* are missing; all these letters contain curves, whereas the five letters listed consist of only straight lines. Therefore *K*, *L*, and *M* (but not *J*) are logical next letters in the pattern.

b. The missing letters—*B*, *C*, *D*, *E*, and *G*—all have the long *e* sound (as do *P*, *T*, *V*, and *Z*). Therefore, the next letters in the sequence would be *K*, *L*, and *M*.

c. <u>O</u>ne, <u>T</u>wo, <u>T</u>hree, <u>F</u>our, <u>F</u>ive, <u>S</u>ix, …

d. SEARCHING FOR PATTERNS IN NATURE.

Note: This pattern is creating by using a code that simply reverses the alphabet:

```
A    B    C    D...
Z    Y    X    W...
```

Exercise #4

Animal Anagrams

LULB	CALEM	NUKKS	SEUMO	BAZER	EAP
BULL	CAMEL	SKUNK	MOUSE	ZEBRA	[APE]

Plant Anagrams

NORC	NOONI	MATOOT	PREPPE	TEBE	EAP
CORN	ONION	TOMATO	PEPPER	BEET	[PEA]

Ape would be the natural pattern response set for the animal group and *pea* for the plant group. That is, students (or dyads) who had discovered that their first five words were all animals (or, conversely, all plants) are more inclined to "see" the word *ape* (or *pea* for the plant group) as the answer to their sixth anagram, even though either answer is correct. Prior conceptions strongly influence how we perceive and conceive new experiences.

Section 3:
Science for All Americans Curriculum Standards

Activity 10

Follow That Star
National Science Education Standards and True North

Expected Outcome
Many learners will have difficulty locating their compass bearings if they are inside a room (especially one that lacks windows that look out on familiar landmarks).

Science Concepts

Earth's magnetic field and its effect on compasses can help people find their bearings relative to a fixed target or direction. (Alternatively, a GPS unit can serve the same function.) The activity also highlights the nature of science (NOS; i.e., reliance on empirical evidence, logical argument, and skeptical review versus simple majority rule).

Science Education Concepts

In terms of financial support and oversight, science education in the United States has historically been the primary responsibility of local school districts and state governments, and only secondarily a national, federal government concern. The 1957 launch of the Russian Sputnik satellite, the 1983 release of the *Nation at Risk* report, and the 2010 work on the Common Core State Standards initiative (see Internet Connections) have each created waves of increased federal involvement in education. Although citizens have in the past resisted the idea of a centrally mandated and assessed national curriculum (i.e., the common practice for most of our international economic partners and competitors), several factors are focusing attention on improved de facto national standards: (1) the dominance of a limited number of textbooks in each science field and grade level and efforts to bring these more in alignment with (2) the *Benchmarks for Science Literacy* (AAAS 1993), *National Science Education Standards* (NRC 1996), and *A Framework for Science Education, Preliminary Public Draft* (NRC 2010) and (3) research on learning in science and articulated K–12 Curriculum-Instruction-Assessment. Additionally, the federal No Child Left Behind Act has now phased in mandated assessments for science as well as reading and mathematics. Teachers need to critically analyze the advantages and limitations of such standards in their daily, unit, and yearlong curricular planning. The *visual participatory analogy* of "standards as a compass" that can help busy teachers keep their bearings and students discover a common destination of big ideas in science is explored.

Follow That Star

Materials

The instructor will need several compasses, an overhead projector compass, and/or a portable GPS unit. A short segment of any of the *Pirates of the Caribbean* movies when Captain Jack Sparrow is using his unusual compass can be used for humorous effect if desired.

Points to Ponder

Science has been seriously retarded by the study of what is not worth knowing, and of what is not knowable.
—Johann Wolfgang von Goethe, German writer and polymath (1749–1832)

Science is organized knowledge … In science the important thing is to modify and change one's ideas as science advances.
—Herbert Spencer, English philosopher and psychologist (1820–1903)

Science is simply common sense at its best—that is, rigidly accurate in observation, and merciless to fallacy in logic.
—Thomas Henry Huxley, English biologist and evolutionist (1825–1895)

"Will you tell me please which way I ought to walk from here?" "That depends a good deal on where you want to get to," said the [Cheshire] cat. "I don't much care where," said Alice. "Then it doesn't much matter which way to walk," said the cat.
—Lewis Carroll, mathematician and author, *Alice's Adventure in Wonderland* (1832–1898)

Procedure

1. Ask the learners to shut their eyes and extend their right arms in the direction they believe represents true north. Make note if anyone points up as north (a likely misconception).

2. While their arms are still extended, ask the learners to open their eyes and note any differences of opinion among their colleagues, think about *empirical evidence* and/or *logical arguments* that support their hypotheses, and change the direction their arms are pointed if they desire. Make a quick assessment as to whether there is a clear majority versus minority view and whether this shifts when the learners look at their peers.

3. Proceed to the separate When Working With Teachers or When Working With Students Debriefing discussions. A segment from one of the *Pirates of the Caribbean* movies can be shown as a humorous transition if desired. Jack Sparrow's compass is not a standard, objective one that points true north regardless of who is holding it, but rather a subjective one that points in the direction of the holder's heart's desire.

Debriefing

When Working With Teachers

Share the science quotes. Then, with this discrepant-event activity as a referent, draw an analogy between finding one's bearings and setting one's course with a compass (or GPS unit) and the efforts to develop, disseminate, and enforce state and national standards for K–12 science education. Initiate a discussion with the following open-ended questions:

1. Is effective science curriculum-instruction-assessment (CIA) just common sense or organized knowledge, as the quotes suggest? Is intelligent CIA defined differently in different countries, and do these definitions change over time? How do differences in curricula and instructional practices contribute to the different results (by countries and states) seen on international science assessments such as TIMSS (Trends in International Mathematics and Science Study) and PISA (Program for International Student

Assessment)? The Internet Connections provide resources for yearlong professional development relative to curriculum standards and standards-based, standardized assessments. See also Activities #21 and #22.

2. The phrase "less is more"—versus "a mile wide and an inch thick" and "overstuffed, but undernourished"—has been used to describe the curricula and assessment results in top-performing countries compared to the United States. Why is it important to determine what is *not* worth knowing in light of developmentally appropriate learning objectives for specific grade-level ranges in science? For example, consider this question: When and at what depth should teachers introduce the various models of the atom (i.e., solid billiard balls → solar system → quantum mechanical)?

3. How might this activity be considered analogous to the curricular perspectives and decisions of individual teachers as limited by the two covers of their textbooks and four walls of their individual classrooms that lack windows to broader, more universal benchmarks? How are national standards for CIA similar to or different from finding true north? Consider the quote from Lewis Carroll.

4. What are the advantages and potential disadvantages of national standards with mandatory assessments?

When Working With Students

Skeptical Review Component of the NOS

1. Are scientific truths determined by majority vote sometimes, always, or never? Can you think of cases in the history of science where well-reasoned commonsense majority views were successfully challenged and overturned by more powerful theories that were initially championed by a minority of scientists (or even just one)?

2. What *empirical evidence* and *logical arguments* support the majority view? What scientific instruments could be used to settle the matter? How do compasses (and/or GPS units) work? Are there any complicating or system-level factors that should be considered because they could change the constancy of compass readings? Are GPS units subject to these same (or different) problems with

stability and reliability of readings? Answering these questions would require hands-on exploratory research and reading about magnets, compasses, and GPS units (see Internet Connections).

Extensions

1. *Magnets Matter and Compasses Count in Science and in History*: Magnets have fascinated natural philosophers and scientists from as early as the philosopher Thales (approximately 600 BC) through their first use in navigational compasses by the Chinese (around 1000 AD) to the scientific explorations of William Gilbert (1600) and studies of electromagnetism by Hans Christian Oersted (1819) and Michael Faraday (1821). Albert Einstein even traced his early interest in science to a compass that was given to him by his father when he was five years old and sick in bed. The history and science of navigation (i.e., means of determining latitude and longitude) can be explored as related to physics and Earth science and as a source of a broadened perception of natural biological diversity and its effect on human history. Related STS resource books include Dava Sobel's *Longitude: The True Story of a Lone Genius Who Solved the Greatest Scientific Problem of His Time*; Amir Aczel's *The Riddle of the Compass: The Invention That Changed the World*; and Jared Diamond's *Guns, Germs, and Steel: The Fates of Human Societies* (biogeography and human history). See also the Internet Connections.

2. *Electromagnetic Connections Case: An STS Study*: The compass played a major role in the discovery of electromagnetism—specifically, Hans Christian Oersted's 1819 discovery that the current flowing through an electric wire created a magnetic field that could deflect a magnet. This initially discrepant event ultimately led to the invention of electric motors, generators (by Michael Faraday and Joseph Henry), and all manners of electrical devices that have transformed modern society and that we take for granted. This story makes for an interesting historical STS case study. See Activities #17, #18, #22, and #23 in *Brain-Powered Science* for demonstration-experiments related to magnetism and electromagnetism.

3. *Geocaching, High-Tech Treasure Hunting, and the Benefits and Burdens of GPS*: Engage students with this game played by adventure seekers who use GPS devices to try to locate containers that are hidden outdoors. GPS applications were originally invented for military applications and have since revolutionized telecommunications, transportation, and other cultural domains. Like all technologies, GPS-based devices hold benefits as well as burdens, risks, and costs (e.g., invasion of privacy and secret government monitoring of citizens). Student teams can complete internet research and have a debate on this science-technology-society (STS) issue. In democracies, informed citizens have a responsibility to help set the direction of how applied research and new technologies are used and regulated in order to maximize benefits and minimize burdens.

Internet Connections

- Disney Educational Productions: *Bill Nye the Science Guy: Magnetism* ($29.99/26 min. DVD): *http://dep.disney.go.com*
- Geocaching—The Official Global GPS Cache Hunt Site: *www.geocaching.com*
- Google Earth: *http://earth.google.com* (free software download)
- How Stuff Works: Compasses: *www.howstuffworks.com/compass.htm*
    GPS Receivers: *http://electronics.howstuffworks.com/gps.htm*
- International and U.S.A. science curriculum standards, assessments, and comparative results:

    Achieve Inc. (independent, bipartisan, nonprofit education reform organization): *www.achieve.org*

    AAAS *Benchmarks*: *www.project2061.org/publications/bsl/online/bolintro.htm*

    College Board Standards for College Success: *Science* and *Mathematics and Statistics* (free PDF downloads): *http://professionals.collegeboard.com/k-12/standards*

    Common Core State Standards Initiative: *www.corestandards.org*

Council of State Science Supervisors' State Science Frameworks (Excel download with links): *www.csss-science.org/frame.shtml*

Education Commission of the States: *www.ecs.org*

International Society for Technology in Education (ISTE): *National Educational Technology Standards*: *www.iste.org/standards.aspx*

International Technology and Engineering Educators Association: *Standards for Technological Literacy: Content for the Study of Technology: www.iteea.org/TAA/Publications/TAA_Publications.html*

National Academy of Sciences' National Research Council: Board on Science Education (*Framework for Science Education*): *http://www7.nationalacademies.org/bose/Standards_Framework_Homepage.html*

National Science Education Standards (1996): *www.nap.edu/readingroom/books/nses/overview.html*

National Assessment of Educational Progress (NAEP): *http://nces.ed.gov/nationsreportcard*

Science Framework for the 2009 National Assessment of Educational Progress (download): *http://nces.ed.gov/nationsreportcard/science/whatmeasure.asp*

Program for International Student Assessment (PISA): *http://nces.ed.gov/surveys/pisa*

See also *www.pisa.oecd.org* and the October 2009 special issue of the *Journal of Research in Science Teaching*: Scientific Literacy and Contexts in PISA Science 46 (8): *http://www3.interscience.wiley.com/journal/122604861/issue*

Trends in International Mathematics and Science Study (TIMSS): *http://nces.ed.gov/timss*

- Interstate New Teacher Assessment and Support Consortium (INTASC): *www.ccsso.org/Resources/Programs/Interstate_Teacher_Assessment_Consortium_(InTASC).html*

- National Board for Professional Teaching Standards (NBPTS): *www.nbpts.org*

Follow That Star

- National Commission on Excellence in Education. 1983. *A nation at risk: The imperative for education reform: www.ed.gov/pubs/NatAtRisk/index.html*
- PhET Interactive Simulations: Magnets and compass: *http://phet.colorado.edu/simulations/sims.php?sim=Magnet_and_Compass*
- University of Virginia Physics Department: Compass Hands-On Explorations (HOEs): *http://galileo.phys.virginia.edu/outreach/8thGradeSOL/MagnetFrm.htm*
- Wake Forest University: Magnetism Videos: 3D Compass and other demonstrations: *www.wfu.edu/physics/demolabs/demos/avimov/bychptr/chptr8_eandm.htm*
- Wikipedia: *http://en.wikipedia.org/wiki*. Search topics: compass, geocaching, geomagnetic reversal, global positioning system, and Hans Christian Oersted

Answers to Questions in Debriefing

When Working With Teachers

1. Whether it is called a K–12 scope and sequence, curriculum map, or learning progression, there is increasing recognition of the need for planned inter-grade-level sequences that develop core science concepts in a rational, scientifically valid manner that presents science as organized knowledge that could become common sense. Most of our international competitors have national standards and assessments.

2. As science advances, the notion of covering every concept becomes increasingly untenable. Deciding what knowledge is the essential foundation for scientifically literate students at different stages in their schooling should enable students to advance to further science studies and informed citizenship. We cannot cover everything if we want to help students "recover" valid conceptual precursors (to build upon), "uncover" misconceptions (to correct), and "discover" the big ideas of science.

3. People are likely to lose their bearings inside a building without orientation relative to the Sun, the North Star (at night), or some other external landmark. Similarly, inside a busy classroom or

school, teachers are likely to have less than optimal bearings with respect to breadth and scope, depth, and within and cross-grade-level sequencing and development of science concepts. If we don't know where we are and where we want to go relative to some valid benchmark, how will we know when we have veered off-course or arrived at our target destination? Simply following the topics at the conceptual level, sequence, and pace established by a textbook may not be optimal. External, research-informed standards can help set a proper direction and trajectory much as a compass (or GPS unit) can do for physical travel. Future research in both science and science education will necessitate modification of currently published standards.

4. Research-informed K–12 curricular goals and learning progressions allow for articulation across grade levels, school districts, and states to better support the conceptual development of both a geographically stable and a mobile population of students. Related standards-based assessments can elicit reliable data that can provide both feedback to the students and their current teachers and information for the next grade-level teacher about students' competencies and learning needs. The risk is that such standards and assessments might result in misguided attempts to micromanage day-to-day instruction and limit teacher creativity and professional judgment by way of scripted, teacher-proof CIA that push out issues of local relevance and anything that isn't tested. Teaching to a poor test is very poor teaching.

When Working With Students

1. There is no single body of representative, elected scientists who rule on scientific truth by a simple majority vote. Throughout the history of science, breakthrough theories have always started as minority viewpoints. Examples of major scientific upheavals include Aristotelian → Newtonian → Einsteinian physics; geocentric versus heliocentric theories in astronomy; phlogiston versus oxidation view of combustion and successive atomic models in chemistry; solid Earth versus continental drift and plate tectonics in geology; and Lamarckian versus Darwinian evolution in biology.

2. A compass or GPS unit could settle the matter; be sure to actually demonstrate this and determine who is right. However, when using compasses, it is important to know that geographic and magnetic north are not identical and have actually flipped over long geologic timescales. Also, compasses can be affected by nearby magnetic fields (including those generated by electric currents) or large masses that contain ferromagnetic metals (i.e., iron, cobalt, or nickel). Compasses depend on the fact that the Earth itself is a huge dipole magnet whose magnetic field is generated by convection of the liquid iron outer core and the Earth's rotation. The source of heat for convection is in turn generated by the radioactive decay of long-lived but unstable isotopes within the crust and mantle. Understanding this series of linked cause-effect relationships requires extended hands-on explorations and targeted readings (see the Internet Connections). The same is true for understanding the science behind GPS; finding and truly understanding one's location using key science concepts as benchmarks takes time.

Activity 11

"Horsing Around"
Curriculum-Instruction-Assessment Problems

Expected Outcome

A manipulative puzzle and a verbal logic challenge (both involving horses) with unexpected-twist-type answers are used to help learners think about the importance of how we define problems and the need to develop a playful, inquisitive orientation to problem solving.

Science Concepts

The nature of science (NOS) involves actively seeking (rather than avoiding) challenging problems that may require looking at situations differently and thinking outside of artificially imposed constraints. Science, like art, is a creative endeavor that involves a sense of purpose, passion, planning, and persistence. Science involves both systematic and serendipitous horsing around with nature's unending puzzles.

Science Education Concepts

These puzzles serve as visual participatory analogies to depict how the way we define and frame problems (or interesting challenges and opportunities) can confine and restrict (or open upon creative, alternative) possibilities for their subsequent resolution. The Points to Ponder quotes suggest that our vision for schools helps set the context for our curriculum-instruction-assessment (CIA) practices. The *Benchmarks for Science Literacy* (AAAS 1993), *National Science Education Standards* (NRC 1996), and *A Framework for Science Education, Preliminary Public Draft* (NRC 2010) can be viewed metaphorically as the instructor or wise philosopher in the two puzzles who help teacher-learners run the race on the right horse, moving in the right direction and at an appropriate speed. Both the current and next-generation of research-informed standards can guide the teacher to focus on core concepts, theories, and habits of mind that will help students construct a solid foundation for understanding, applying, and continuing to learn science.

Materials

Photocopies of the Two Horses and Two Riders (manipulative) puzzle, the Winning a Horse Race by Being the Slowest (verbal logic) puzzle, the "Schools We Envision" quote on page 133, and the Riding a Dead Horse parody (Extensions, p. 138)

"Horsing Around"

Points to Ponder

The solution which I am urging, is to eradicate the fatal disconnection of subjects which kills the vitality of our modern curriculum. There is only one subject-matter for education, and that is Life in all of its manifestations. Instead of this single unity, we offer children—Algebra, from which nothing follows; Geometry, from which nothing follows; Science, from which nothing follows; History, from which nothing follows, a Couple of Languages, never mastered; and lastly, most dreary of all, Literature … Can such a list be said to represent Life, as it is known in the midst of living it? The best that can be said of it is, that it is a rapid table of contents which a Deity might run over in His mind when he was thinking of creating a world and had not yet determined how to put it together.

—Alfred North Whitehead, English mathematician and philosopher (1861–1947) in his 1916 essay "The Aims of Education" (see Internet Connections)

The schools we envision are exciting places: thoughtful, reflective, engaging and engaged. They are places where meaning is made. They are places that resemble workshops, studios, galleries, theaters, studies, laboratories, field research sites, and newsrooms. Their spirit is one of shared inquiry. The students in these schools feel supported in taking risks and thinking independently. They are engaged in initiating and assessing their ideas and products, developing a disciplined respect for their own work and the work of others. Their teachers function more like coaches, mentors, wise advisors, and guides than as information transmitters or gatekeepers. They offer high standards with high levels of support, creating a bridge between challenging curriculum goals and students' unique needs, talents, and learning styles. They are continually learning because they teach in schools where everyone would be glad to be a student, or a teacher—where everyone would want to be—or could be—both...

—*Learner-Centered Curriculum and Assessment for New York State* (New York State Education Department 1994)

Procedure

Begin the activity by distributing photocopies or projecting an image of the "schools we envision" quote. Initiate a brief, open-ended conversation by asking the teachers (or students) these questions:

- Does this quote accurately represent your personal experiences in schools?
- Would you like to be a teacher (or student) in such a school?
- What are you willing to contribute or invest to help start to close the gap between this ideal vision and the current reality?

Suggest to the learners that problem resolution in science, school, and life depends on the clarity with which we perceive and subsequently define or frame problems in the context of broader "systems of parts ←→ whole relationships." Encourage a sense of playfulness and creative flexibility in thinking about problems related to learning and teaching in schools by inviting the learners to horse around with the following puzzles.

Puzzle #1: Two Horses and Two Riders Puzzle

Ask the learners to make a crease between sections A and B (the left and right sides) of the puzzle, bend it back and forth several times, and neatly separate the two pieces by tearing along the straight crease. Challenge learners to arrange the two pieces in such a way that the two riders are properly seated on two horses. Request that they initially play with this puzzle on their own and avoid looking to peers for an answer. If learners discover one solution, challenge them to develop a second solution and decide which is more elegant. Let the learners continue to "horse around" with this manipulative puzzle while you give them puzzle #2, the horse and rider verbal logic puzzle.

Historical note: This classic manipulative puzzle dates back to Sam Loyd's Trick Mules puzzle of 1871 (which likely was based on a series of even older puzzles that are pictured and discussed at *www.threehares.net/puzzles.html*). When showman P.T. Barnum came across Loyd's puzzle, he reportedly bought it for $10,000. He subsequently distributed it to millions of people who visited his circus, making it one of the most popular puzzles of all time. The horse and rider version was later reprinted in *Scientific American* in the late 1890s. How the mind generates creative insight to solve such problems received early attention in the following publications:

- Bartlett, F. C. 1951. *The mind at work and play.* Chapter 6. Boston, MA: Beacon Press.
- Scheerer, M., K. Goldstein, and E. G. Boring. 1941. A demonstration of insight: The horse-and-rider puzzle. *American Journal of Psychology* 54 (3): 437–438.

Puzzle #2: Winning a Horse Race by Being the Slowest

Read or project the following verbal logic puzzle (without the embedded answers):

An old king who is nearing his death decides he will bequeath his entire intact kingdom to only one of his twin sons. He decides that he will make his decision based on a horse race between his sons. Given that wise deliberation and patience are important in a leader, he decides to change the normal rules for horse races so that the son who owns the slower horse will become the new ruler of the entire kingdom. Ask the learners: What practical dilemma does this unusually designed race present for resolving this important decision?

To help visualize the dynamics of the two horses and two riders system, invite an assistant up to the front of the room to play-act being one of the two brothers riding his horse with you. This visual hint should make it evident that each of the twin brothers would fear that the other would cheat by intentionally having his horse go slower than it is capable of running. Given the disincentive for going faster, both brothers would slow down their horses to allow the other to charge ahead and win the race (which really means losing the kingdom in this "slower horse wins" race). Thus, the race would eventually come to a complete stop.

Realizing they have what seems to be a no-win problem, the two brothers travel up a tall mountain to seek a wise philosopher's advice. Pondering the problem from a different perspective, the philosopher tells them in two words how to make sure the race comes to a conclusion that is both fair and timely. What are the two words spoken by the philosopher? Alternatively, if the philosopher could only use one

word what might it be? As with the first puzzle, let the learners individually "horse around" with this puzzle for a while before discussing the solutions and follow-up questions (see Debriefing).

Debriefing

Puzzle #1: When Working With Teachers and Students

After reviewing and critiquing various solutions, use the manipulative puzzle as a *visual participatory analogy* to initiate an open-ended discussion that addresses questions such as these: What are some particular cases of practices and policies in science classrooms or broader school communities where you believe we are "riding the wrong horse"? How does the way we define the "problem" of learning or teaching science set constraints on the possible solutions? The history of science offers numerous examples where insights and discoveries occurred when a scientist looked at what others had previously seen from a different angle or new perspective. Consider this truism: "A problem well defined is half solved."

Also note that different kinds of problems draw on different mental abilities (or multiple intelligences). Manipulating physical materials, as with this first puzzle, employs both visual-spatial and body-kinesthetic intelligences, unlike the second puzzle, which relies more on verbal-linguistic, logical-mathematical intelligences and less on visual spatial intelligences (although the play-acting draws on the latter). How can teachers reframe the system of curriculum-instruction-assessment to better activate, assess, and develop a broader range of mental abilities? How can students learn to draw on their mental strengths to help overcome their weaknesses?

Puzzle #2: When Working With Teachers

After citing the solution ("change horses"), discuss how this verbal logic puzzle is a *visual participatory analogy* for many of our current science curriculum-instruction-assessment policies and practices. One might consider the ever-increasing size of textbooks and curriculum guides as a call to race through and cover (albeit superficially) material at an ever-faster rate to have students (horses) compete well against others in the school, district, state, or nation. By contrast, research on learning, the *National Science Education Standards*, the *Benchmarks for Science Literacy*, and the national curricula of our international peers all call for covering fewer core concepts and skills in more depth (less is more) in carefully articulated scope and sequences (or multigrade learning progressions). Perhaps we need to slow down, change horses, and cooperate. Share and compare and contrast the 1916 Whitehead quote, the current nature of curriculum-instruction-assessment (CIA), and the vision of the NYSED quote.

When Working With Students
Challenge students to consider the following questions:

1. What current personal study habits do you need to change to help you learn science more efficiently? How can "slow and steady win the day" (i.e., distributed versus massed practice)?

2. Is learning more about racing against others or about doing better than your own past performance and extending the limits of your current understandings?

3. How can you help both your teacher and peers establish a cooperative learning environment where all succeed and there are no losers?

Extensions
When Working With Teachers
1. *Nursery Rhyme and Reason*: "For the want of a nail the shoe was lost. For the want of a shoe the horse was lost. For the want of a horse the rider was lost. For the want of the rider the battle was lost. For the want of a battle the kingdom was lost. And all for the want of a horseshoe nail."

    This nursery rhyme, which has existed in many variations for centuries, is a nice entry into a discussion of root-cause analysis (see Internet Connections) as a problem-solving technique to trace the cause-and-effect trail from the end failure back to the root cause. The failure of the American school to provide equitable, excellent science education for *all* students is a multifaceted systems problem with boundaries that extend well beyond the

school grounds. However, the collective power of schools driven by a compelling vision of becoming learning organizations is often overlooked. Perhaps teachers cooperatively sharing best-practice curriculum-instruction-assessment ideas and resources on a regular basis could be the collective horseshoe nail that saves the kingdom. Perhaps in contrast to the model of interschool and interdistrict athletic competitions, teachers should look to professional collaboration that extends well beyond the boundaries of their classrooms, schools, and districts. The notion that we are all in this together is also captured nicely in the *Chicken Soup and Mousetrap* parable circulated on the internet (see Internet Connections).

2. *Riding a Dead Horse*: *An Educational Policy and Practices Parody*: Extend the "horsing around" analogy one step further by reading and discussing the parody described in the 1999 book *If You're Riding a Horse and It Dies, Get Off* by Jim Grant and Char Forsten (alternatively, find the widely circulated, noncopyrighted text in the Internet Connections). Refining and reforming practices in a way that truly transforms teaching and turns schools into the type of learning organizations described in the vision quote are not easy undertakings. However, transformation can begin with one teacher defining a problem in a different way and working in concert with another teacher. Shared laughter can be considered a lubricant that helps overcome friction, cognitive inertia, and other impediments to growth and change among learners of all ages.

When Working With Students

Throughout the science course, teachers should regularly ask students to solve challenging problems to teach them how to develop a playful, creative, problem-solving mindset. A wide variety of books of such problems are available, although they tend to focus more on the physical sciences than on life sciences (see Bibliography).

A Bibliography of Creative, Problem-Solving Science Books for Grades 9–12 Students

- Adams, S. M. 2002. *Chemistry puzzles and games.* Greensburg, PA: Games for Science Education.

"Horsing Around"

- Brecher, E., and M. Gerrard. 1997. *Challenging science puzzles.* New York: Sterling. This book contains conceptual and qualitative physics quandaries (with solutions) covering heat, light, and sound (20); space (15); household physics (22); general phenomena (41); and tongue-in-cheek physics (8). Many of these puzzles can be translated into discrepant-event demonstrations.
- Delorenzo, R. A. 1992. *Problem solving in general chemistry.* 2nd ed. Dubuque, IA: William C. Brown. The author integrates quantitative problem solving and real-life applications of chemistry.
- Epstein, L. C. 2000. *Thinking physics: Understandable practical reality.* San Francisco, CA: Insight Press. This book contains 582 page-long, cartoon-like puzzles with multiple-choice answers to questions on mechanics, fluids, vibration, light, electricity, magnetism, relativity, and quanta (many of which can be used as discrepant-event demonstrations).
- Gardner, M. 1978. *Aha! Insight.* New York: Scientific American/ W.H. Freeman.
- Gardner, M. 1982. *Aha! Gotcha.* New York: Scientific American/W. H. Freeman. See also other science, math, and logic and puzzle books by this author.
- Jargodzki, C., and F. Potter. 2000. *Mad about physics: Braintwisters, paradoxes, and curiosities.* Hoboken, NJ: John Wiley & Sons. This book contains 397 puzzles (with answers) on real-world (and often peculiar) applications of basic physics principles. Along the margins are humorous quotations by and about scientists. See also by the same authors *Mad About Modern Physics* (2004).
- Jargodzki, C. P. 1976. *Science brain-twisters, paradoxes and fallacies.* New York: Charles Scribner's Sons.
- Jargodzki, C. P. 1983. *More science brain-twisters and paradoxes.* New York: Van Nostrand Reinhold. Contains more than 160 common-sense-eluding science questions with answers.
- Jewett, J. W. 1994. *Physics begins with an M ... Mysteries, magic and myth.* Boston, MA: Allyn and Bacon. See also *Physics Begins With Another M* (1995). Both books combine real-world application-type questions, discrepant-event demonstrations, and misconceptions.

- Moscovitz, I., and I. Stewart. 2006. *The big book of brain games: 1000 playthinks of art, mathematics, and science.* New York: Workman Publishing. See other books by same authors.
- Shortz, W., ed. 1991. *The giant book of games.* New York: Times Books/Random House. This book contains more than 200 visual, word, and mathematical games, puzzles, and mind teasers selected from *Games* magazine. See also *2nd Giant Book of Games* (1996).
- Walker, J. 2006. *The flying circus of physics with answers.* 2nd ed. New York: John Wiley & Sons. This book contains more than 700 intriguing and fun, real-world questions with answers and references.

Internet Connections

- Chicken Soup and Mousetrap parable: *www.wow4u.com/mousetrap/index.html* and *http://message.snopes.com/showthread.php?t=14050*

- Horse and Rider Puzzle (posted in multiple locations): *http://goodreasonblog.blogspot.com/2006/03/horse-and-rider-puzzle.html*

- SmartKit Puzzle Playground: Online Puzzles and Brain Games: See interactive flash version of Horse and Rider puzzle: *www.smart-kit.com/s846/horse-riding-puzzle*

- International and U.S.A. Science Curriculum Standards: See listing in Activity #10 (pp. 125–127).

- Riding a Dead Horse: An Educational Policy and Practices Parody:

    A to Z Home's Cool: *http://homeschooling.gomilpitas.com/humor/056.htm*

    Business Performance Improvement Consultancy: *www.bpic.co.uk/articles/deadhorse.htm*

    Native American Wisdom: *www.mississippiwebsite.com/humor.htm*

    Thought You Should Know News: *www.tysknews.com/LiteStuff/riding_a_dead_horse.htm*

- Root Cause Analysis as a Problem-Solving Technique:

    process.nasa.gov/documents/RootCauseAnalysis.pdf

www.youtube.com/watch?v=GOVeO5_0qD0&feature=related (Titanic sinking)

http://en.wikipedia.org/wiki/Root_cause_analysis

- *The Aims of Education*: 1916 Mathematical Association presidential address and the first essay in a 1929 book of the same name by Alfred North Whitehead: *www.ditext.com/whitehead/aims.html*

Answers to Questions in Procedure

Puzzle #1 Solution

The curvature and strange proportions of the two horses on section A and the bend of their necks relative to the portion of their manes visible on the two riders both point to the solution. That is, section B (the riders) needs to be laid perpendicular just below the head of one horse and the hind legs of the other horse. Two "new" horses appear in the correct proportion and orientation relative to the riders. An alternative creative solution that almost works is to make a crease between the two horses to form an upside-down *V* and make a small *W*-shaped crease between the two riders to rest on top of the upside-down *V*. (However, careful inspection reveals that this clever 3-D solution is not quite right.)

Puzzle #2: Two-Word Solution

Change horses. Each brother should ride the other brother's horse and make it go as fast as possible so that the horse he owns "loses" and he wins by virtue of "owning" the slowest horse.

Puzzle #2: One-Word Solution

Cooperate or *share*. The two brothers could agree that the winner of a fair race would *share* his good fortune with the other brother and press their father to understand that two heads are better than one.

Magical Signs of Science
"Basic Indicators" for Student Inquiry

Expected Outcome
An invisible, colorless message becomes colored and legible when it is sprayed with colorless household ammonia.

Science Concepts

The chemistry of acid-base indicators; the solubility, volatility, and reactivity of ammonia gas; and the nature of science (NOS) are explored. Extensions include consumer chemistry challenges and an STS case study in which students can apply and extend these ideas.

Science Education Concepts

Advertisements capture consumers' attention and motivate them to buy certain products. Similarly, teachers need to sell science to students by walking the talk of research-informed best practices. Magic signs with (dis)appearing messages (or color-changing letters) provide both a visual participatory analogy for and concrete example of the kind of interactive curriculum-instruction-assessment practices called for by standards documents (e.g., AAAS's *Benchmarks* and the NRC's *National Science Education Standards* and *A Framework for Science Education, Preliminary Public Draft*). These documents advocate for minds-on CIA practices that develop latent student interest in, engagement with, and understanding of science concepts. Alternatively, this activity can serve as an analogy for the use of discrepant events to make visible hidden preinstructional conceptions that students may hold about a given concept. These include misconceptions that are incompatible with accepted scientific views and are quite tenacious if not directly activated and challenged by teacher-introduced anomalies (see Appendix A, p. 295).

Materials

- Poster-size sheet of newsprint or nonglossy flip chart paper for the sign
- Solution of an acid-base indicator as ink
- Low concentration of a weak acid or base (depending on the specific indicator used) in a transparent spray bottle as the "developer"

Different combinations of indicator (ink) and developer (spray) will produce different colors. Phenolphthalein "ink" is colorless in its acid form and invisible when dry. When sprayed with colorless household ammonia (or even ammonia-based window cleaners, if they

Magical Signs of Science

are not colored), phenolphthalein ink turns pink. Phenolphthalein was used for decades as the active ingredient in laxative tablets but was removed in the 1990s due to health risks. It is still the most common indicator used in schools (where ingestion is not an issue). Other possible chemical combinations are described in Extension #1. If desired, they can be used to make part of the sign visible at the beginning of the demonstration. *Optional*: Any song that features

Safety Note
The instructor should wear safety goggles and the learners should be two meters away from the sign.

Points to Ponder

When Science from Creation's face
  Enchantment's veil withdraws,
What lovely visions yield their place
  To cold material laws!

—Thomas Campbell, English poet (1777–1884)

Sweet is the lore which Nature brings;
  Our meddling intellect
Misshapes the beauteous form of things:—
  We murder to dissect.

—William Wordsworth, English poet (1770–1850)

The creative use of imagination is not only the fountain of all inspiration in poetry and art, but it is also the source of discovery in science, and indeed supplies the initial impulse to all development and progress. It is the creative use of imagination which has inspired and guided all the great discoveries in science.

—Sir William Higgins, British astronomer (1824–1910)

color (e.g., "The Rainbow Connection," as performed by Kermit the Frog from the Muppets).

Procedure

1. At least one hour before class, prepare one or more magic signs in the following fashion:
   a. Spray water-absorbent newsprint or flip chart paper lightly with water and let it dry. (This makes the paper a little crinkly and easier to paint on without ink runs.)
   b. Use a paintbrush to paint a hidden message with a solution of phenolphthalein indicator ("ink") in large block letters that will be visible to learners in the back of the room; let it dry a little before hanging it to avoid ink runs, and be sure to mark the paper so you know how to orient it for later use.
   c. Put the colorless household ammonia into an unmarked, colorless, clear spray bottle for use as the "developing solution."
   d. When the message is dry, hang the seemingly blank poster on a wall. Here are some sample hidden messages: *Misconceptions Matter*, *Science is not secondary. It's elementary and FUNdaMENTAL!* and *Science = empirical evidence + logical argument + skeptical review*.

2. Begin by inviting the learners to read the take-home message or learning objectives for the day from the blank sign. When learners react with blank stares, tell them that learning is an act of individual cognitive construction that requires effort. But interactions with peers and real-world phenomena, as well as targeted assistance from a teacher, can help students develop conceptual understanding. Learning is both an individual and cooperative team sport.

3. Hold up the spray bottle of household ammonia (or substitute a colorless, ammonia-based window cleaner) and ask learners to describe what they see. They should mention that you have a clear, colorless solution of unknown chemical composition that—based on the data they have—may or may not be water. With some flair, spray the sign with the household ammonia mist. The hidden message will appear in vivid pink block letters.

Safety Note

Although this is the same product used in some window cleaners, the instructor may wish to wear goggles. Also, make sure the learners are at least 2 m away from spray zone and that the room has good ventilation.

4. Ask the learners for additional observations; those nearest to the sign should recognize the characteristic ammonia smell (*empirical evidence*). Based on this observation, ask learners to develop a theory (*logical argument*) that explains the "magical" appearance of the hidden words.

5. As the learners do this, the sign will begin the transition back to its colorless form. Ask the learners the following questions:
   a. How would a chemist explain the sign changing from colorless to pink and back to colorless?
   b. What kind of liquid "developer" would result in a less transient, more lasting pink color?

Note: The big idea of *constancy and change* can be modeled, as the dry sign can be reused many times if ammonia is used as the developing spray.

Debriefing

When Working With Teachers

After leading the teachers through the inquiry process, compare and contrast the quotes and discuss how good science teaching enhances a sense of appreciation, wonder, and awe about the nature of our world. Both basic and applied science depend on disciplined creativity as much as any of the humanities, and the stories science tells are every bit as engaging as popular literature. Consider the Harry Potter series by J.K. Rowling that has captured the attention of everyone from middle schoolers to senior citizens. Each of the seven books introduces a new Defense of the Dark Arts teacher (who, along with the potions teacher, is the most chemistry-like teacher in the series). The fifth book, *Harry Potter and the Order of the Phoenix* (2003), introduces Professor Umbridge (pp. 238–245). The passage on these pages is a powerful parody of a decontextualized hands-off and minds-off approach to curriculum-instruction-assessment ("wands away and quills out … a return to basic principles"). Share the passage (or the scene from the movie) and discuss how such a CIA approach can take the magic out of even a Hogwarts class.

One can also draw an analogy between the hidden message and preinstructional conceptions that students hold about an array of

natural phenomena and the role of discrepant events in activating, assessing, and challenging science preconceptions. Also consider discussing the pedagogical and practical advantages of using household chemicals and easily reusable setups: They are typically cheaper and safer, send a message that chemistry is everywhere, and are easy to replicate within and across classes.

When Working With Students

This activity can be used either as the Engage phase (do not provide premature answers before allowing students to explore the questions) or in the Elaborate phase of the 5E Teaching Cycle on acids, bases, and indicators. In either case, the activity calls for additional experimentation (see the Extensions and Internet Connections). See also Activity #18 for a biology-related use of this chemical reaction.

Extensions

1. *Consumer Chemistry Challenge*: How would using other common household acids, bases, and indicators create different kinds of magical signs? Diluted alkaline-spray solutions made with solid, nonvolatile, water-soluble, stronger bases such as sodium hydroxide (lye) will cause a more permanent color change unless a second acid spray of sufficient strength and concentration is used to reverse the reaction. Other indicators could be used to produce different color signs. For example, thymolphthalein changes from its colorless form (below pH 9.3) to a blue form in bases (above pH 10.5). It is commonly sold in toy stores as Blue Magic Ink that disappears when it is exposed to air (due to the presence of acid-forming carbon dioxide). Red cabbage juice, a natural universal indicator, goes through a whole rainbow spectrum of colors at different pHs. Turmeric (the yellow spice in curry dishes) dissolved in alcohol changes from yellow to red in the presence of a base. Water or alcohol extracts from a variety of other colored fruits, vegetables, and flowers change color one or more times across the pH range of 1–14. Several chemical supply companies (e.g., Educational Innovations [*www.teachersource.com*] and Steve Spangler Science [*www.stevespanglerscience.com*]) sell an older formulation of goldenrod 8.5 in. × 11 in. paper that

turns a bright ("blood") red when it comes in contact with bases such as household ammonia, baking soda, or baking powder. Acid solutions such as vinegar, lemon juice, or, with time, the carbon-dioxide in the air will reverse the dramatic color change and allow the paper to be reused. Indicators that change color in the pH range of 6–8.5 (e.g., bromthymol blue and phenol red) can be used in a weakly basic solution as a "voice-activated reaction" that will change color after being "talked to" (exposed to the acid-forming carbon dioxide gas). See L. R. Summerlin and J. L. Ealy's *Chemical Demonstrations*, Vol. 1, page 40 (ACS Press 1985) and Purdue University in the Internet Connections.

Beyond acid-base indicators, other types of invisible and color-changing inks include invisible citric acid or lemon juice ink that can be developed with heat from a flame, hot plate, or heat gun; spray laundry starch that can be diluted in a little water to use as invisible ink that will change to a blue-violet color when sprayed with a water bottle that has had several tablespoons of tincture of iodine added to it; and sunscreen lotions with a SPF of 15 or more that can be used as invisible ink on white typing paper and will appear when the paper is illuminated with a black or UV light.

2. *A Science-Technology-Society (STS) Historical Case Study*: Ammonia, like other top 10 industrial chemicals (see the American Chemical Society's *Chemical & Engineering News* annual issue that discusses these chemicals), has a fascinating history that reflects the varied ethical uses to which basic science can be put via derived technologies. Ammonia gas was first isolated by Joseph Priestley in 1774 and called alkaline air. It is a critical chemical ingredient used in the manufacture of both explosives and fertilizers. Prior to the work of the German Jewish chemist Fritz Haber (and his collaborator, Carl Bosch), "fixed nitrogen" was only available from naturally occurring, unevenly distributed nitrate compounds (e.g., Chilean sodium nitrate found in guano or bird manure). Chemists did not know how to duplicate the ability of lightning and symbiotic bacteria (that live in the root nodules of legumes) to break the strong triple bonds in diatomic nitrogen gas molecules. The Haber-Bosch process for converting abundant atmospheric nitrogen into ammonia played a pivotal role

in WWI by allowing Germany to be able to wage an extended war. Since then, it has multiplied agricultural productivity and reduced world hunger. Fritz Haber also played a central role in developing other aspects of chemical warfare. Ironically, he was subsequently forced to flee anti-Semitic Nazi Germany (see Internet Connections: Wikipedia). The book *The Alchemy of Air* by Thomas Hager (2008) tells the fascinating story of the Haber-Bosch collaboration that fueled the rise of Hitler but later fed the world. Similarly, the case of Alfred Nobel and the positive and negative uses of dynamite presents another interesting example of how a chemical invention can be used for varied purposes, with divergent moral implications.

Internet Connections

- About Chemistry

    Acid-Base Indicators, their pH ranges, and respective colors: *http://chemistry.about.com/library/weekly/aa112201a.htm*

    Chemistry Experiments and Demonstrations You Can Do at Home (e.g., make disappearing ink): *http://chemistry.about.com/od/homeexperiments*

- Biology Corner: Scientific Method: Select: Carbon Dioxide Production (respiration lab): *www.biologycorner.com/lesson-plans/scientific-method*

- Cell Biology Animations (see pH and water): *www.johnkyrk.com*

- Chem Guide: Indicators: *www.chemguide.co.uk/physical/acidbaseeqia/indicators.html*

- Disney Educational Productions: *Bill Nye the Science Guy: Chemical Reactions* ($29.99/26 min. DVD): *http://dep.disney.go.com*

- Doing Chemistry: Movies of chemical demonstrations: Colorful reactions of red cabbage juice: *http://chemmovies.unl.edu/chemistry/dochem/DoChem098.html*. Cabbage juice indicator: *http://chemmovies.unl.edu/chemistry/dochem/DoChem099.html*

- Fun Science Gallery: Experiments with Acids and Bases (red cabbage): *www.funsci.com/fun3_en/acids/acids.htm*

Magical Signs of Science

- General Chemistry Online: Indicators and their color changes/ pH ranges: *http://antoine.frostburg.edu/chem/senese/101/acidbase/indicators.shtml*
- Goldenrod Paper as an Acid-Base Indicator

    Becker Demonstrations: *http://chemmovies.unl.edu/chemistry/beckerdemos/BD022.html*

    Daryl's Science: *www.darylscience.com/Demos/Goldenrod.html*

    Science Hobbyist: *http://amasci.com/amateur/gold.html*

    Steve Spangler Science: *www.stevespanglerscience.com/experiment/00000040*

- International and USA Science Curriculum Standards: See listing in Activity #10 (pp. 125–127).
- PhET Interactive Simulations: *http://phet.colorado.edu/simulations/sims.php?sim=pH_Scale*

    http://phet.colorado.edu/en/simulation/acid-base-solutions

- Purdue University: Voice-activated chemical reaction: *www.chem.purdue.edu/bcce/Voice_Activated_Reaction.pdf*
- Steve Spangler Science: Do you Have Acid Breath?: *www.stevespangler.com/experiments*
- University of Virginia Physics Dept: Soda Science, Flower Indicators, and Acid-Base Tea Party:

    http://galileo.phys.virginia.edu/outreach/8thGradeSOL/SodaScienceFrm.htm

    http://galileo.phys.virginia.edu/outreach/8thGradeSOL/FlowerIndicatorsFrm.htm

    http://galileo.phys.virginia.edu/outreach/8thGradeSOL/AcidBaseFrm.htm

- Wikipedia: *http://en.wikipedia.org/wiki*. Search topics: acid rain, ammonia, Fritz Haber, Haber process, and pH indicators (includes list)

Answers to Questions in Procedure, steps #4 and #5

4. Depending on their prior experiences with acid-base chemistry, the learners may or may not recognize that the *system* consists of an acid-base indicator that turns from colorless to pink in the presence of a base. Phenolphthalein ($C_{20}H_{14}O_4$) was used as invisible ink and ammonia was used as the developer. As an interesting aside, phenolphthalein was used for decades as the key ingredient in Ex-Lax but was replaced with other chemical laxatives during the 1990s due to concerns about possible carcinogenicity. There is no known health hazard for its use in titrations and other experiments where it is not ingested.

5. a. The transition from pink back to colorless can be explained in the following way.

    The fact that household ammonia can be smelled at a distance from the liquid indicates that ammonia is either a volatile liquid or a water-soluble gas whose molecules can diffuse from the spray site to the human observer's nose. Our olfactory sensory system is our portable chemical identification kit that evolved to alert our primate ancestors to potential dangers, food, mates, and more. Ammonia is in fact a water-soluble gas that reacts with phenolphthalein and changes it to its pink alkaline form at pH 9. As the water evaporates from the wet sign, some of the dissolved ammonia gas also escapes to the atmosphere. Additionally, carbon dioxide in the air reacts with the water to form a weak carbonic acid solution that converts the indicator back to its colorless, invisible acid form. Acid-base chemistry depends on reversible reactions whose equilibrium state can be changed by temperature, pressure, and the presence of other chemicals.

    b. Developing solutions of nonvolatile, solid bases (e.g., alkali or alkaline Earth hydoxides) would lead to a more permanent color change. Use of solutions of such strong bases in this demonstration would require safety goggles, gloves, a lab coat, and a great distance from students.

Verifying Vexing Volumes
"Can Be as Easy as Pi" Mathematics

Expected Outcomes

Two sheets of 8.5 in. × 11 in. paper are used to construct two cylinders with surprisingly different volumes (as related to the diameter differences between the unmentioned, missing bases). Extensions include the empirical discovery of pi and a volume illusion with two cone-shaped glasses.

Science Concepts

These activities feature the mathematical formula for the volume of a cylinder, the importance of quantitative measurements versus simple "eyeball" estimates, and the advantages of measuring in metric versus English units.

Science Education Concepts

Mathematics is both a key to unlocking the secrets of nature and, for many students, a stumbling block to conceptual understanding in science. Conceptual, limited-math approaches to science teaching notwithstanding, all science teachers need to find ways to integrate mathematics instruction in science (and mathematics teachers would do well to integrate real-world science examples in their instruction as well). The AAAS Project 2061's *Benchmarks for Science Literacy* (AAAS 1993) devotes an entire chapter to the mathematical world (see Internet Connections). Most science textbooks have a unit on measuring and quantitative skills. But in too many cases, students are asked to practice these skills in non-inquiry-oriented, decontextualized contexts. This activity demonstrates that mathematics can answer interesting questions via FUNdaMENTAL hands-on play and minds-on work. Science curriculum-instruction-assessment (CIA) standards call for synergistic coordination with the teaching of mathematical concepts and skills.

Materials

- 2 sheets of 8.5 in. × 11 in. paper and a calculator for each dyad
- Tape (or paper clips), plus 2 sheets of 8.5 in. × 11 in. transparency film
- Box of rice, cereal, or dried beans
- 2 cone-shaped glasses (Pilsner glasses or champagne flutes) and food coloring are used in the Extension.

Safety Note

No food should be consumed in a laboratory.

Verifying Vexing Volumes

Points to Ponder

For he who knows not mathematics cannot know any other science; what is more, he cannot discover his own ignorance, or find its proper remedy ... Mathematics is both the door and the key to the sciences.
—Roger Bacon, English philosopher and clergyman (1220–1292)

No human investigation can be called scientific if it cannot be demonstrated mathematically.
—Leonardo da Vinci, Italian polymath (1452–1519)

The Universe is a grand book of philosophy. The book lies continually open to man's gaze, yet none can hope to comprehend it, who has not first mastered the language and characters in which it has been written. This language is mathematics; these characters are triangles, circles, and other geometric figures.
—Galileo Galilei, Italian astronomer and physicist (1564–1642)

Procedure

1. Model Building: Give each dyad two standard 8.5 in. × 11 in. pieces of paper and ask them to build two different-shape cylinders or cans (do *not* mention the two "missing" end pieces) that use an entire sheet of the paper. Secure the two shapes with either tape or paper clips.

2. Ask dyads to discuss the following questions:
   a. What is the same (versus different) about these two objects?
   b. If you wanted to use a container that could hold the greatest volume (e.g., packing water bottles for a trip into the desert), would the volume of these two cylinders be the same or different? Develop a *logical argument* to support your educated guess. (*Note:* The instructor should not cite the importance of

the relative diameters and surface areas of the unmentioned, missing circular end caps or bases.)

3. Investigation or *skeptical review*: Challenge learners to do the following:

   a. Design an *empirical* method of determining the relative volumes of the two cylinders. After the learners have proposed (and, if time permits, tested out) various solutions, "spill the beans" with the following silent teacher demonstration. Construct two cylinders using the clear transparency film sheets and clear tape to connect the ends. Place the shorter 8.5 in. cylinder inside a larger glass container (e.g., a 2 L beaker) and insert the 11 in. tall cylinder inside the shorter but wider cylinder. Fill the tall, narrow cylinder with dried beans, rice, or breakfast cereal. Pull up the inner cylinder and remove it to allow the beans to fall into the outer, shorter cylinder (which will not fill completely).

   b. Develop a mathematical model or logical explanation that would predict the answer (and/or explain the results) of your empirical test. (*Note:* r^2 plays a disproportionate role versus the height.)

Debriefing

When Working With Teachers

Discuss the quotes and the following open-ended questions: Were you a student who was comfortable, confident, and competent with mathematics? What can you do to improve your students' attitudes about and abilities in mathematics? How can graphing calculators, spreadsheets, and computer simulations be used in a way that helps students truly understand mathematics rather than merely getting the right answers? Should students still be taught to do order-of-magnitude estimations and other mental math in their heads? What has this activity taught you about teaching as telling and learning as listening, and mindless memorization versus teaching as engaging and challenging students to discover and understand meaningful patterns? How can effective teaching be considered both a science and a performing art (Tauber and Sargent Mester 2007)? Should science teachers try to create magical moments in class?

When Working With Students

Share the Points to Ponder quotes about the relationship between science and mathematics, and discuss these questions:

1. What are the advantages of making measurements and calculations in metric versus English units? (*Note:* 1 cm^3 = 1 ml versus the relationship between in.^{3} and oz.)

2. How do the empirical results compare to your initial guesstimate?

3. Does the mathematical argument help you make sense of the empirical results? Does understanding the science and mathematics take away the magic of the experience? Note that students are unlikely to initially consider the disproportionate effect of the different radii for the missing end caps of the cylinders relative to the effect of height on volume.

4. Archimedes makes a useful historical case study of a man ahead of his times. This Greek mathematician, scientist, and engineer (287–212 BC) creatively integrated the use of measurements, mathematics, and model building to make important discoveries and inventions centuries before these approaches and ideas became commonplace in science (see Activity #14, Extension #4). Galileo's (1564–1642) rediscovery of the writings of Archimedes helped launch the scientific revolution and re-establish mathematics as a critical tool. Although mathematics has been more commonly associated with the physical sciences, the biological sciences (e.g., proteomics, genomics, population biology, ecological modeling, and biotechnology) are becoming increasingly mathematical in their approaches.

Extensions

1. *Easy as Pi*: Challenge students to rediscover the approximate value for pi by measuring the circumference and diameter of various cans (or circles) of different sizes. Again, this is much easier done with metric units than with the English system.

    a. Does there appear to be a mathematical relationship that connects these two variables?

b. What would a graph of circumference versus diameter look like?
c. Assuming that this relationship also holds true for three-dimensional spheres, how can one calculate the diameter of a sphere without cutting the sphere in half to measure it?
d. Tasty Mathematical Joke: What is the volume of a cylinder with radius z and height a?

Students also can explore the structural strength of a cylinder (or column) versus other shapes that can be built with the same piece of paper and its role in Greek and Roman architecture. As a complement to hands-on explorations, students can systematically play with mathematical simulations and explore Archimedes's groundbreaking work (see Activity #14).

2. *Magic Glasses: Two for One, One for All*: A classic magician's trick depends on the effect of increasing radii on volume. Obtain two tall, identical, cone-shape glasses. Prior to the demonstration (and out of sight from your audience), fill one glass to the brim with colored water and then carefully pour half of that volume into the second glass (so that the two glasses have equal volumes). Alternatively, wait to color the water with food coloring until after you divide the volume in half, then use different colors for the two halves, so that, when combined, they will result in a third color (i.e., red + yellow = orange, yellow + blue = green, or red + blue = violet). Show the two glasses of water to the learners and ask them if all the water will fit into one glass. As the water lines will be well above the halfway height of the glasses (due to the smaller radius of the cone-shape glass at the bottom versus progressively up the glass to the top), most people will say no. Demonstrate the optical and mathematical illusion by pouring all the water into one glass. Relate this illusion back to the two-cylinder activity. Note that the larger the difference between the radius at the top and bottom of the glass and the taller the glasses, the more dramatic the effect. Cheap plastic champagne glasses (or flutes) work for hands-on exploration of this FUNomena.

3. *More Metric Measurement Merriment*: Metric measuring skills can also be developed by challenging students to check their visual

Verifying Vexing Volumes

estimates of a variety of optical illusions (e.g., are these two lines the same or different lengths?). Google "optical illusions" or see the extensive listing of websites in *Brain-Powered Science*, Activity #5. Students also can explore mathematical principles such as the Pythagorean theorem ($a^2 + b^2 = c^2$ for right triangles) and the area of a circle (see the MacTutor, University of Toronto, and Wikipedia websites in Internet Connections).

Internet Connections

- AAAS *Benchmarks*: *www.project2061.org/publications/bsl/online/bolintro.htm*
- Archimedes: Biographical Information: See websites in Activity #14, Internet Connections.
- International and USA Science Curriculum Standards: See websites in Activity #10, Internet Connections.
- Java Applets on Mathematics: *www.walter-fendt.de/m14e*

    Select: Circumference of a Circle (interactive computer simulation)
- MacTutor History of Mathematics: Pythagoras of Samos: *http://www-history.mcs.st-and.ac.uk/Biographies/Pythagoras.html*
- Optical Illusions (sample sites):

    www.michaelbach.de/ot

    www.optillusions.com

    www.planetperplex.com/en/optical_illusions.html
- PBS/NOVA: *www.pbs.org/wgbh/nova* and *http://shop.wgbh.org*

    See DVD biographies (typically $19.95) and free online teacher guides on various mathematician-scientists, such as *Infinite Secrets*: *The Genius of Archimedes, Galileo's Battle for the Heavens,* and *Newton's Dark Secrets.*
- University of Toronto, Department of Physics: Physics Flash Animations: Miscellaneous: Area of a Circle as a Limit: *www.upscale.utoronto.ca/GeneralInterest/Harrison/Flash*
- Wikipedia: Pythagoras of Samos: *http://en.wikipedia.org/wiki/Pythagoras*

Activity 13

Answers to Questions in Procecure, steps #2 and #3

2. a. The surface area of the side walls, not counting the "missing" circular end caps, are the same; the volumes are different.

3. b.

   English → Metric Conversion (for a standard 8.5 in. × 11 in. piece of paper)

   8.5 in. × 2.54 cm/in. = 21.59 cm = circumference of tall, narrow cylinder

   11 in. × 2.54 cm/in. = 27.94 cm = circumference of short, wide cylinder

Formulas

Pi = π = ~ 3.14 (can be empirically discovered, see the Extension)
Circumference = π × diameter = $2\pi r$ or Radius = $C/(2\pi)$
Area of a circle = πr^2
Volume of a cylinder: area of base × height = $\pi r^2 \times h$

Calculations

Radius = circumference/(2π)
Tall, narrow cylinder: r = 3.438 cm = 21.59/(2 × 3.14)
Short, wide cylinder: r = 4.449 cm = 27.94/(2 × 3.14)

Area of circular base × Height = Volume of cylinder

Tall, narrow cylinder: 3.14 × (3.438 cm)^2 × 27.94 cm = 1,037 cm^3
Short, wide cylinder: 3.14 × (4.449 cm)^2 × 21.59 cm = 1,342 cm^3
(29% larger)

Note the ratio of volumes is 1,342/1,037 = 1.29 = 11/8.5 = ratio of heights of the 2 cylinders

Extension #1, a–d

A plot of circumference versus diameter would be a line with a slope ($\Delta C/\Delta d$) = 3.14 = pi. The relationship of C = 3.14 (d) can be used to calculate the diameter of a sphere without cutting it by simply measuring its circumference. If V = $\pi r^2 h$ and we substitute *pi* for π, the letter *z* for *r*, and the letter *a* for *h*, we get V = piz^{2}a. Since z^2 = z·z, V = pizza.

Activity 14

Archimedes, the Syracuse (Sicily) Scientist
Science Rules Balance and Bathtub Basics

Expected Outcome

Learners are guided to discover simple, quantitative relationships that support their qualitative observations of playing on seesaws (with different-size children) and taking baths.

Science Concepts

The law of levers and the principles of water displacement and buoyancy point to the role of quantitative measurements and mathematics in the sciences. Repeating Archimedes's historic experiments provides a meaningful, real-world entry into empirical sciences (e.g., use of balances and graduated cylinders). The law of levers also demonstrates the law of conservation of energy (energy input = energy output). More generally, mathematical models support, complement, and extend the use of physical and conceptual models in science.

Science Education Concepts

Science is a discipline built on empirical data, logical arguments, and skeptical review. National standards documents (i.e., *National Science Education Standards* and *Benchmarks for Science Literacy*) emphasize the importance of the history and philosophy of science and the nature of science. They also call for their integration into curriculum-instruction-assessment (CIA) units that are designed around a carefully planned scope and sequence of minds-on activities that include (but are not limited to) hands-on explorations. The history of science includes a number of classic experiments that can be rediscovered by students to simultaneously teach basic laboratory skills and concepts rather than mindless, process-only laboratory exercises. The concept of leverage is a powerful metaphor in many contexts. For instance, within science education, both discrepant-event activities and analogies leverage students' natural curiosity to help them lift the heavy weight of big ideas that have evolved over hundreds of years. Similarly, science and mathematics provide mutual leverage for meaningful student learning and eureka-type insights.

Materials

- 1 or more simple (commercial or homemade), equal arm math balances that have hooks at regular intervals from which to hang standard mass weights

Archimedes, the Syracuse (Sicily) Scientist

- *Activities for Learning, RightStart Mathematics*: Math Balance (R7) for $20. 888-775-6284. *www.activitiesforlearning.com/index.asp?PageAction=VIEWPROD&ProdID=53*
- EAI Education: Math Balance With Weights: #532306/$13.95: 1-800-770-8010. *www.eaieducation.com/ProductInfo.aspx?productid=532306.* A 26 in. × 8.75 in., durable plastic balance includes 20, 10 g weights, self-adhesive labels, and an instruction booklet.
- Beakers (or cut-off 2 L plastic soda bottles or plastic tennis cans)
- Pre- and post-1982 pennies
- Various objects that are denser than water
- Graduated cylinders

Points to Ponder

Give me a firm spot on which to stand and I will move the earth.

—Archimedes, Greek scientist, mathematician, engineer, and natural philosopher (287–212 BC)

With accurate experiment and observation to work upon, imagination becomes the architect of physical theory.

—John Tyndall, Irish physicist (1820–1893)

Th_*_ difficulty in science is often not so m_*_ch how to make the discove_*_y but rath_*_r to _*_now that one has m_*_de it.

—John D. Bernal, English X-ray crystallographer

* *Mini-Discovery Activity:* Fill in the six missing letters to complete the quote and spell out a word that is commonly used to describe the "aha, I got it" experience in science (eureka).

Procedure

The following activities can be used as either a teacher demonstration or hands-on exploration, depending on the time and materials available and teaching context.

Part #1: Seesaw Science From the Syracuse (Sicily) Scientist

Ask the following questions:

1. If two children want to balance each other on a seesaw (or teeter-totter), what experimental factors or variables need to be considered?

2. If the two children are of unequal weights, what can they do to counterbalance the weight difference so they can still balance? Elicit a range of ideas beyond getting another smaller friend to join them on the light side of the seesaw. After eliciting their ideas, either demonstrate how to counterbalance the weight difference with an equal arm math balance or have small groups of learners directly explore how to achieve balance in the case of equal versus unequal weights.

3. What is the quantitative relationship between the weights of the two children and their distance from the fulcrum? Use the math balance and standard-mass hanging weights to empirically discover the mathematical relationship.

4. Archimedes, the ancient Greek polymath, discovered the law of levers (and pulleys) and demonstrated their use when he reputedly single-handedly moved a ship to dry dock using a compound pulley system. Did his discovery enable him to do less work (i.e., use less energy)?

Part #2: A Historical Hypothesis Challenge Story: The "Gold Standard" of Science

King Hiero II of Syracuse (in Sicily) commissioned an artisan to construct a crown of gold for him. After giving the artisan a specific mass of gold, the king had reason to doubt the artisan's honesty even though the finished crown weighed the same as the original mass of

Archimedes, the Syracuse (Sicily) Scientist

gold he had been given. The king turned to his "technical advisor," Archimedes, to resolve the matter. The challenge for Archimedes was to determine a nondestructive way to figure out if the crown was pure gold (or if it had been diluted with cheaper metals, such as silver or copper, either via plating or mixing with another metal to form a uniform gold-colored alloy) without cutting or melting the crown down, thereby damaging it. Temporarily at a loss as to how he should proceed, Archimedes had a sudden "aha" insight when he stepped into a bath. Overcome with intellectual excitement, he ran through the streets completely naked shouting "Eureka!" (I've got it!) Ask the following questions about his discovery:

1. What scientific principles did Archimedes discover? Demonstrate (or have learners directly explore) water displacement with objects of different sizes, shapes, and densities.

2. How did he use this insight to prove that the artisan was dishonest?

3. How do empirical data, logical arguments, and skeptical review form the "gold standard" of science as a discipline? Science teachers should be able to answer these questions by using a mini case study from the history of science. Students will need access to reference materials such as the Internet and hands-on explorations. (*Note:* The December 8, 2006, issue of *Scientific American* [see Internet Connections] and other sources question the historical accuracy of this classic story.)

Debriefing

When Working With Teachers

Briefly share the unifying themes of systems, models, evidence and explanation, change and constancy, measurement and scale, and form and function as found in the *National Science Education Standards* (NRC 1996) and *Benchmarks* (AAAS 1993). Discuss how these themes relate to the phenomena of levers and buoyancy. How can we incorporate more opportunities for students to see that science is built on empirical data, logical arguments, and skeptical review (versus blind acceptance of authority and faith in the textbook and teacher)? Use this activity as a *visual participatory analogy* to discuss how integrating

science and mathematics provides mutual leverage points to move students away from preconceptions (see Driver et al. 1994, chapter 12, "Water/Floating and Sinking," and chapter 21, "Forces," for an overview related to these activities); add a motivational real-world component to student learning; and create opportunities for students to have eureka-type insights. Also discuss how historical vignettes, case studies, and laboratory re-enactments can be used to help students gain insight into the history and nature of science.

When Working With Students

These activities would need to be further developed as hands-on explorations to provide a great real-world context for learning and applying the skills of making linear, volume, and mass measurements. As a follow-up to Part #1, students could use spring scales to examine a number of first-class lever-type tools to see how they work. For Part #2, students can perform standard experiments involving volume measurement of regular versus irregular-shape objects via water displacement. The volume of regularly shaped objects can be calculated via a formula (e.g., $l \times w \times h$ for a rectangular solid; other formulas for other shapes) and compared to that obtained by water displacement to establish the validity of the latter method for use with irregularly shaped objects. Alternatively, clay or playdough can be formed into regular versus irregular solids and the solid's volume can be compared via the two methods. Additionally, students can compare the mass-versus-volume plots for different volumes of water versus cooking oil to determine the constancy and uniqueness of the density of pure substances (i.e., the slope of a mass versus volume plot = density). Depending on the grade level, you may scaffold students' understanding of mass measurements by progressing from an equal arm balance to a triple beam balance to a digital electronic balance. Also, in addition to hands-on explorations, students can systematically play with the Java Applets on Physics computer simulations (see Internet Connections).

Extensions

1. *Bathtub Basics*: Challenge students to consider the following scenario and questions:

Archimedes, the Syracuse (Sicily) Scientist

If a loaded boat were resting in a sealed, water-filled dry dock and its load of metallic ore were dumped into the water surrounding the boat,

   a. Would the water level rise, fall, or stay the same?
   b. How could you build a simple model system to empirically answer this question?

2. *Making Sense of Pennies*: Examine pre- and post-1982 pennies.

   a. How are they the same or different? *Hint:* If the pre-1982 pennies represent the amount of pure gold that King Hiero II had given to the Syracuse artisan and the post-1982 pennies represent the crown, how would Archimedes have been able to prove easily that the crown was definitely fake (not made of 100% pure gold)?
   b. What does your answer imply about the composition of the newer pennies, and how could you test your hypothesis?
   c. How would a shipload of old pennies compare to a shipload of new pennies if the two otherwise identical boats were floated in identical, adjacent water-filled dry docks? Consider both where the respective water lines would be and how high the water would rise in the docks.

3. *Penny for Your Thoughts Teacher Demonstrations*: Pre- and post-1982 pennies can be chemically distinguished from one another by using a file to make a few notches around the perimeter of the coins (to expose the zinc interior on the newer pennies) and placing them in separate test tubes of concentrated hydrochloric or sulfuric acid. The nearly pure copper pre-1982 penny will not react, whereas in the case of the post-1982 penny the acid will chemically react with the exposed zinc to form an aqueous salt and release hydrogen gas that will eventually cause the intact exterior copper shell to float to the top of the test tube. Alternatively, a mixture of the pre- and post-1982 pennies can also be used as a mass-related analogy for isotopes of the same element.

4. *Explore the History of a Genius*: Archimedes also invented the water screw, made an accurate estimation of the value of pi, deduced formulas for the volumes of three-dimensional (3-D) shapes, and came close to inventing calculus more than 2,200

Safety Note

Do this demonstration only in a laboratory with an eye wash available. Wear indirectly vented chemical splash goggles, gloves, and an apron and do NOT use nitric acid, as it reacts with copper to form a green solution of aqueous copper nitrate and harmful, orange-colored nitrogen dioxide gas fumes.

years ago (centuries before Isaac Newton)! Unfortunately, his untimely death at the hands of a Roman soldier slowed the evolution of empirical science until Galileo (1564–1642) rediscovered his lost writings. A case study of Archimedes can be set in the broader context of the history and philosophy of science (including the nature of science). See the AAAS *Benchmarks*, Archimedes, NRC *National Science Education Standards*, PBS, and Wikipedia websites in Internet Connections. Useful books include the following:

- Bendick, J. 1995. *Archimedes and the door of science.* Bathgate, ND: Bethlehem Books/Ignatius Press. This 142-page book is written for students in grades 4 and higher.
- Hakim, J. 2004. *The story of science: Aristotle leads the way.* Washington, DC: Smithsonian Books and NSTA Press. Volume 1 of the three-part series includes an extended discussion of Archimedes's work (chapter 17, pp. 146–159). This series is a great resource for both teachers and students.
- Matthews, M. R. 1994. *Science teaching: The role of history and philosophy of science.* New York: Routledge. This is a scholarly but accessible book for teachers who wish to probe deeper.

Internet Connections

See also the Internet Connections related to density in Activity #3.

- American Association for the Advancement of Science (AAAS). 1993. *Benchmarks for science literacy.* www.project2061.org/publications/bsl/online/bolintro.htm.

    See chapter 1 ("Nature of Science"), chapter 2 ("Nature of Mathematics"), and chapter 10 ("Historical Perspectives").

- Archimedes: Biographical Information: discoveries, inventions, surviving works, and more:

    www.cs.drexel.edu/~crorres/Archimedes/contents.html

    www.gap-system.org/~history/Mathematicians/Archimedes.html

    www.math.nyu.edu/~crorres/Archimedes/contents.html

    http://www-personal.umich.edu/~lpt/archimedes.htm (Gold Thief & Buoyancy story)

Archimedes, the Syracuse (Sicily) Scientist

www.scientificamerican.com/article.cfm?id=fact-or-fiction-archimede

http://www-history.mcs.st-andrews.ac.uk/Biographies/Archimedes.html

- Contextual Science Teaching (History of Science resources): *http://sci-ed.org*
- Disney Educational Productions: *Bill Nye the Science Guy: Buoyancy* ($29.99/26 min. DVD): *http://dep.disney.go.com*
- History of Science Society: *www.hssonline.org/main_pg.html*
- HyperPhysics, Department of Physics and Astronomy, Georgia State University:

    Mechanics: Fluids: Buoyancy and Archimedes Principle: Explanation and Concept Maps: *http://hyperphysics.phy-astr.gsu.edu/hbase/hframe.html*

- Illinois State University Physics and Astronomy Lecture Demonstrations: Select Mechanics—Hydrostatics: *http://learning.physics.iastate.edu/DemoRoom/Index.htm#*
- International Society for the History and Philosophy of Science: *www.hopos.org*
- Int'l History, Philosophy and Science Teaching Group (SHiPS Resource Center): *http://www1.umn.edu/ships/hpst*
- *Internet History of Science Sourcebook*: contains numerous links that cut across ancient to modern times: *www.fordham.edu/halsall/science/sciencesbook.html*
- Java Applets on Physics: *www.walter-fendt.de/ph14e/index.html*

    Select: lever principle, Newton's second law, and buoyant force in liquids

- National Research Council (NRC). 1996. *National science education standards*: *www.nap.edu/readingroom/books/nses/overview.html*.

    See Content Standard G: History and Nature of Science, pp. 170–171 (grades 5–8) and pp. 200–204 (grades 9–12) and Program Standard C, pp. 214–218 (coordination with mathematics).

- Museum of Science and Industry-Chicago: See How to Build a Lever: *www.msichicago.org/online-science/activities*

- PBS/NOVA: *www.pbs.org/wgbh/nova* and *http://shop.wgbh.org*
  See the episode: *Infinite Secrets: The Genius of Archimedes.*
- University of Iowa Physics and Astronomy Lecture Demonstrations (video clips): Heat and Fluids: *http://faraday.physics.uiowa.edu* (5 demonstrations on density and buoyancy)
- University of Michigan: Physics Demonstration Catalog: Archimedes Principle: *http://webapps.lsa.umich.edu/physics/demolab/Content/demo.aspx?id=19*
- University of Virginia Physics Department: Levers, Compound Machines and Mechanical Advantage:

  http://galileo.phys.virginia.edu/outreach/8thGradeSOL/compoundmachineFrm.htm

  http://galileo.phys.virginia.edu/outreach/8thGradeSOL/LeversFrm.htm
- Wake Forest University: Newton's Laws Videos: 14 hook balance (F1 × d1 = F2 × d2): *www.wfu.edu/physics/demolabs/demos/avimov/bychptr/chptr2_newton.htm*
- WikiPedia: *http://en.wikipedia.org/wiki*. Search topics: Archimedes and the history and philosophy of science

Answers to Questions in Procedure

Part #1: Seesaw Science From the Syracuse (Sicily) Scientist

1. The relative weight of the two children and their relative distances from the fulcrum

2. The lighter-weight child should sit farther away from the center than the heavier child. Some seesaws allow the position of the fulcrum to be shifted away from the center of the lever arm so that the lighter child has a longer lever arm without shifting position on the seat.

3. The law of levers states that $F_1 \times d_1 = F_2 \times d_2$, where F is the force due to weight and d is the distance from the fulcrum for child #1

Archimedes, the Syracuse (Sicily) Scientist

and child #2. Thus, there is an inverse relationship between force and distance in creating leverage.

4. No, energy = work = force × distance. Technically, the relevant distance in this energy-equals-work equation is the upward (or downward) motion of the lever arm at the point that the upward (or downward) force is applied and not the distances of the two lever arms in the equation from #3. Thus, when a lower-weight person balances out a heavier person by sitting farther back, the lower weight person moves a proportionally greater distance up and down compared to the heavier person. The distance up or down in which the two weights move can be measured if desired to show that in the absence of significant friction, an equal arm balance demonstrates the law of conservation of energy or that the energy input is equal to the energy output. A lever, a pulley, and other simple machines allow work to be done more easily, often by exerting a smaller force over a longer distance (e.g., using a screwdriver as a lever to pry off the lid of a paint can). Simple machines do not save us work as defined by physicists. Electric, gas, or other fuel-powered machines supplement or replace human-powered work with another source of energy.

Part #2: A Historical Hypothesis Challenge Story: The "Gold Standard" of Science

1. The scientific principles discovered by Archimedes include the following:
   a. Water displacement: A submerged object will displace a volume of water equal to its own volume.
   b. Buoyancy: A floating object will displace a volume of water equal to its mass.
   c. Density: Different materials have different mass-to-volume ratios that can be used to identify pure substances.

2. Archimedes proved the artisan was dishonest by noting that a mass of gold equal to that originally given to the artisan displaced a smaller volume of water than the crown displaced. He deduced that the artisan had used a lower-density, cheaper metal as a substitute for some of the original gold. To ensure the overall weight was the same, the artisan had to use a larger volume of

the lower-density metal, making the final volume of the crown greater than it would have been if it had been made of pure gold. This is an example of "gold standard" scientific reasoning.

3. *Empirical data, logical arguments,* and *skeptical review* collectively are the "gold standard" that distinguishes science as a field of study. This should be a yearlong theme that cuts across all units of study in a science classroom at all grade levels.

Answers to Extensions

1. *Bathtub Basics*: A model of a dry dock can be built using a uniform-bore, cylindrically shaped container of water (such as a 3- or 4-ball plastic tennis can or a 2 L plastic soda bottle with the top cut off). The cylinder is partially filled with water, a small plastic cup (boat) is loaded with pennies or other metal weights to the point of near sinking, and a marker is used to place a line on the outside of the container to note the height of the water level. When the boat is emptied, the water level is observed to fall because the weights take up less volume when they are submerged than when they are floating as part of the boat's cargo.

2. *Making Sense of Pennies*: The composition of pennies was changed midyear during the 1982 mint year because the value of the copper in the coins (i.e., if illegally melted down and sold as scrap) was worth more than a penny! Pre-1982 pennies are nearly pure copper (95%) and have a uniform composition throughout. To avoid the scarcity and rationing or conservation mindset (reminiscent of the 1943 WWII pennies with carbon steel inside and zinc outside), post-1982 pennies were made to look like the real thing by placing a thin copper exterior over a nearly pure zinc interior (97.5% zinc, 2.5% copper). Zinc was chosen as a cheaper yet durable alternative, and the new pennies were kept the same size as the old. Because zinc has a lower density than copper (7.13 versus 8.96 g/cm^3), newer pennies weigh less (by about 0.5 g, near the limit of hand detectability) than older pennies and will displace a smaller volume of water if they are loaded into a small boat. This can be modeled using the previously described setup. With rising metal and minting costs, the newer, cheaper pennies once again cost the U.S. government more than a penny to make ($.0167 as of November 2007). Thus, there have been repeated calls to phase out pennies altogether.

Measurements and Molecules Matter

Less Is More and Curriculum "Survival of the Fittest"

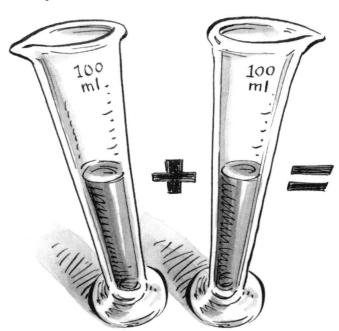

$$50.0 \, ml + 50.0 \, ml = ?$$

Expected Outcome

The volume nonadditivity that occurs when combining equal volumes of alcohol and water (e.g., 50.0 ml + 50.0 ml = 96.0–97.0 ml of the mixture) leads to a consideration of the importance of experimenter skill, integrity, and honesty; the "significance" of significant digits; and the atomic/kinetic molecular theory of matter.

Science Concepts

Measurement precision, significant digits (or figures), and the atomic/kinetic molecular theory are all explored. A volume reduction occurs when mixing the two different liquids because of the different sizes and shapes of the ethanol versus water molecules; their intra- and intermolecular forces; and their corresponding, ever-shifting intermolecular "holes." Mass is conserved because atoms are not created or destroyed, but volume is not conserved because the empty spaces between molecules can be reduced via mixing.

Science Education Concepts

An experiment where "more becomes less" is used as *a visual participatory analogy* for the converse idea that less is more with respect to curriculum that has a more focused coverage. The *Benchmarks for Science Literacy* (AAAS 1993), the *National Science Education Standards* (NRC 1996), and *A Framework for Science Education, Preliminary Public Draft* (NRC 2010) argue for uncovering relevant preconceptions and discovering in depth a smaller number of common themes, unifying concepts, or core and cross-cutting big ideas. This research-informed recommendation is in contrast to teaching that attempts to cover a large number of less important concepts, which results in superficial learning that cannot later be recovered from memory and creatively applied by learners (who "miss the forest for the trees"). Also, laboratory inquiry skills are best learned in a meaningful context (i.e., where measurements matter) rather than as decontextualized skills to be used later in a course. Meaningful, exploration-type laboratory investigations that are both hands-on and minds-on are powerful forms of curriculum-embedded, formative assessments. They also create a need to know that can be resolved in subsequent Explain-phase learning activities.

Materials

Teacher Demonstration
- As a student-assisted, interactive teacher demonstration-experiment, this discrepant-event activity requires a minimum of 100 ml of standard laboratory-grade ethanol (or methanol).

Measurements and Molecules Matter

Drugstore rubbing alcohol (either ethanol or isopropanol) will only work if the concentration is in approximately the same range. The 50–70% rubbing alcohol varieties will not result in as noticeable a volume reduction because they are already diluted with water.
- About 50 ml of BBs and an equivalent volume of marbles are needed as simple, pourable models of water and ethanol molecules.
- Space-filling molecular models of an ethanol molecule and a water molecule can be purchased from science supply companies or can be made using proportionally sized Styrofoam balls from a craft store (1 in. diameter for hydrogen atoms, 2 in. for oxygen atoms, and 2.5 in. for carbon atoms).
- Several 100 ml and 10 ml graduated cylinders are also needed.

Safety Note

Denatured ethanol [95%] is poisonous to drink but safe to handle if it is kept away from flames or sparks that could ignite it. If this activity is done as a hands-on experiment, students should wear safety goggles, and adequate room ventilation is important.

Hands-On Exploration Option

Depending on the age and maturity of the students and lab facilities and safety equipment (i.e., safety goggles), the teacher may want to do this activity as a modified hands-on exploration. Different teams could be assigned to complete one of the following volume-measurement tasks: adding water to water, adding ethanol to ethanol, or adding ethanol to water. This approach will keep the amount of ethanol-water mixture that needs to be disposed of to a minimum, and it will automatically produce discrepant results across the different teams. Each pair of students will need approximately 50 ml of each of the two liquids they are assigned and two 100 ml graduated cylinders.

Optional: 10-Minute University: The World's Fastest-Talking Man Teaches the World's Greatest Lessons. (2004). An illustrated book and CD by Jim Becker, Andy Mayer, Bob Tzudiker, Noni White, and Mark Brewer (illustrator), with John "Mighty Mouth" Moschitta (CD "lecturer"). New York: Barnes & Noble Books.

Points to Ponder

It is the weight not the number of experiments that is to be regarded.
—Isaac Newton, English physicist and mathematician (1642–1727)

The experiment serves two purposes, often independent one from the other: it allows the observation of new facts, hitherto either unsuspected, or not yet well defined; and it determines whether a working hypothesis fits the world of observable facts.
—Rene J. DuBos, French American bacteriologist (1901–1982)

If in some cataclysm, all scientific knowledge were to be destroyed, and only one sentence passed on to the next generation of creatures, what statement would contain the most information in the fewest words? I believe it is the atomic hypothesis (or atomic fact, or whatever else you wish to call it) that all things are made of atoms—little particles that move around in perpetual motion, attracting each other when they are a little distance apart, but repelling upon being squeezed into one another. In that one sentence, you will see, there is an enormous amount of information about the world, if just a little imagination and thinking are applied.
—Richard Feynman in *Six Easy Pieces: Essential Physics by Its Most Brilliant Teacher* (1996)

Thus every discipline is like a hologram ... Holographic logic was anticipated two and a half centuries ago by William Blake in a simple image from "Auguries of Innocence," where he suggests that we can "see a World in a Grain of Sand." Every academic discipline has such "grains of sand" through which its world can be seen. So why do we keep dumping truckloads of sand on our students, blinding them to the whole, instead of lifting up a grain so they can learn to see for themselves? Why do we keep trying to cover the field when we can honor the stuff of the discipline more profoundly be teaching less of it a deeper level?
—Parker J. Palmer in *The Courage to Teach* (1998, p. 125)

Measurements and Molecules Matter

Procedure

1. For both teachers and students, mention that the keys to understanding science are learning the core concepts, how they interrelate, and how they apply to real-world contexts, as well as learning how to think scientifically (including how we know what we know in science). Share the quotes from Newton and DuBos as an introduction to laboratory-based learning.

2. The following steps are written for an interactive teacher demonstration experiment with volunteer assistants to help make the measurements and mixtures.

    a. Use two 100 ml graduated cylinders to separately measure out as close to 50 ml of liquid #1 (ethanol) and 50 ml of liquid #2 (water) as possible. Volumes in the range of 49.6–50.0 ml are fine, but it is important to not exceed 50.0 ml in either graduated cylinder. The identity of the two liquids can be shared or left as unknown. If desired, an optional drop of food coloring added to one or both of the liquids provides an easy visual indicator of when the two liquids are thoroughly mixed in the next step (e.g., red + blue = purple or yellow + blue = green; see also Extension #2 (p. 182) for a related variation on density). Alternatively, the result may be even more discrepant if both clear, colorless liquids appear to be the same (although the difference in smell between the two liquids may be noticed by the assistants). In either case, ask the assistants to take their unmixed liquids to several classmates to verify the precise volumes of the unmixed liquids. Or, if a document camera is available, focus on the meniscus so everyone in the class can simultaneously see the volumes. Ask: Does it matter if we record the results as 50, 50.0, or 50.00 ml, or are these numeric values all the same? What does the term *precision* mean, and how does it relate to the particular measuring device used? Be sure to point out that the meniscus curves downward in both liquids, and volumes should be read with the bottom of the curve at eye level. If the volumes suggested by the learners are less precise, ask them to recheck their figures. If the volume of either of the two liquids exceeds 50.0 ml, ask students to tilt the graduated cylinders sideways and use a long plastic eyedropper to remove and discard a little of the liquid from one or both of their graduated cylinders before mixing them together in the next step.

b. Carefully pour (without spilling or splashing) liquid #1 into the graduated cylinder of liquid #2 and back again into graduated cylinder #1 so that the two are combined and thoroughly mixed in one graduated cylinder. Ask the learners to note any observations that accompany the mixing and think how these might help account for the final volume of the mixture.

c. To help learners better visualize the approximately 3–4 ml reduction in volume, place the same volume of water that was lost into several 10 ml graduated cylinders to pass around the room. Brainstorm possible explanations for this volume reduction. Does the reduction in volume imply that matter was also lost? Why or why not? How could this be checked or controlled for in the experimental design?

d. As time permits, collectively discuss how to check and control for various possible experimental factors. For example, repeat the experiment by adding together two graduated cylinders of 50.0 ml of liquid #1 and then two cylinders of liquid #2. In both cases, the combined volumes obtained when mixing the pure liquids with themselves should be within 0.5 ml of the algebraic sum (100 ml). Invite other possible explanations to account for how more has become less when adding two different liquids but not when adding pure liquids together. Whether or not the idea of different-size intermolecular holes is mentioned, move on to the next step.

e. Perform a silent demonstration of adding 50 ml of BBs to 50 ml of marbles, with a slight shaking to cause mixing. Elicit ideas as to how this simple macroscopic model might relate to the nonadditivity of liquid mixtures versus the volume additivity of pure liquids. See if anyone thinks he or she knows the identity of the two originally clear, colorless liquids. Identify the liquids as ethanol and water. Challenge the learners to discuss the strengths and weaknesses of the BB-and-marble model and the space-filling molecular model. Hold up larger, more realistic-looking space-filling (not ball-and-stick) models of each molecule to show their relative

Measurements and Molecules Matter

size and shape (or distribute kits so each group can construct their own models of H_2O and CH_3CH_2OH). Molecular animations can also be used to show students a more dynamic molecular scale perspective (see Internet Connections: Cell Biology Animations).

f. Optional Challenge Question: How might we explain a situation where two different liquids were physically mixed together (without a chemical reaction) and the volume of the mixture was observed to be greater than their additive sum?

Debriefing

When Working With Teachers

Ask how many of them had previously done experiments (as students) that showed that significant figures are truly significant in science or who were provided a compelling yet simple example of the need for a particulate (or molecular) theory of matter. Discuss how in attempting to cover more and more content (and perhaps even complete a set number of lab exercises), we are less likely to teach in inquiry-oriented ways that uncover the nature of science (i.e., its foundation on empirical evidence, logical argument, and skeptical review); promote an ability to creatively transfer lessons learned to new contexts; or help students construct meaningful learning that lasts (and can be recovered from memory and used) long beyond a test. Students are typically asked to accept truly far-out, outrageous ideas such as atoms and molecules on faith on the authority of the teacher or textbook (e.g., see Activity #2). Consider, for instance, that when you hold 18 ml (or 18 g or 1 mole) of water in your hand, scientists believe you are actually holding 6×10^{23} incredibly tiny, V- or Mickey Mouse–shape H_2O molecules. Each molecule is in a state of constant translational, rotational, and vibrational motion (including the internal movements of their three constituent atoms)!

If available, use one of the one-minute audio mini-course summary lectures from *10-Minute University* (see p. 175) as a humorous way to point to the challenge of covering an ever-increasing number of science concepts in one year. Alternatively (or as a complementary prop), hold up a copy of a large science textbook and ask if it is possible to cover all of the book's information in one year. As

an option, consider projecting all the end-of-chapter terms—such as from a biology chapter on fungi—to point out that not even all science (or even biology) teachers know (or need to know) all these specialized terms.

Share the quote from *The Courage to Teach* (p. 176) as a lead-in for having the teachers work in groups of four to develop quick preliminary lists of curriculum exclusion principles. These principles or criteria could be used to either eliminate or restrict the time devoted to concepts that are identified as being less important or relevant so that more time is available to spend on big ideas. Compare the teachers' lists of criteria to the following list and to their experiences with Activity #15.

A concept should be *excluded* from (or given restricted time in) a curriculum if it meets the following criteria:

- The only or primary justification for inclusion is that it will be needed for the next course. (*conceptual relevance*)
- It cannot be made understandable (*cognitive accessibility*), given
    - students' cognitive abilities (*developmental appropriateness*)
    - available lab, demonstration, and multimedia facilities and equipment (*demonstrability*)
- A relevant connection cannot be made between the concept and the personal or social environments of the students' lives. (*life applicability, utility, and relevance*)
- It will not play a role in subsequent discussions. (*disciplinary centrality*)
- Its inclusion consistently has been found to confuse and frustrate students and/or dampen enthusiasm for the discipline. (*meaningfulness and motivational appeal*)
- Other more central, FUNdaMENTAL concepts better meet the inclusion criteria implied above, and curriculum time is a factor. ("*survival of the fittest*")

Some teachers might feel this set of criteria would submit traditional curriculum and textbooks to a slash-and-burn-type campaign. Actually, it is more of a call for teachers and curriculum designers to discover the relevance and power of core disciplinary ideas as called for by the *National Science Education Standards* (NRC 1996, p. 109),

Measurements and Molecules Matter

the *Benchmarks for Science Literacy* (AAAS 1993), and the *Framework for Science Education, Preliminary Public Draft* (NRC 2010, pp. 1–14). In any case, deciding what concepts *not* to teach (or spend only limited amounts of time on) is an important corollary to deciding what to teach, and these decisions need to be informed by local- and state-level standards and assessments rather than driven by what topics happen to be in the textbook.

When Working With Students

Activity #18 can be used to emphasize the importance of faithfully reporting actual results, rather than what you think should happen (or other lab groups' reports). Many famous scientific discoveries have been made when an experimenter noticed something unusual or a mistake and followed up on the serendipitous discrepancy rather than ignoring it as others had done. This activity also serves as an engaging introduction to significant digits and the need for careful measurements as a means of making interesting discoveries. This approach stands in contrast to skills-only labs where students practice for future, potentially more interesting uses. Finally, the activity is one of many (see the Extensions below and Activity #8 in this book, as well as Activity #20 in *Brain-Powered Science*) that can be used to convince students that the atomic and kinetic molecular theory really is a sensible (i.e., its effects can be experienced through the senses and are logical) and powerful explanatory and predictive model for a wide variety of macroscopic events. From a molecular viewpoint, all matter is "holey"—that is, nothing is truly solid.

Extensions

1. *Molecular Mixed Drinks and Density Demonstration*: The previous experiment can be combined with an experiment on density and (im)miscibility by setting up two contrasting large test tubes or graduated cylinders. Fill the first tube approximately halfway with water colored with yellow food coloring; then carefully run approximately the same volume of blue ethanol down the side of the tube to layer it gently on top of the yellow water. This setup up will be relatively stable, as the denser water is on the bottom. Contrast this to the second tube, where you reverse the

order of the liquids by putting the less-dense blue alcohol in first, then attempt to layer the more-dense yellow water on top. In the latter case, mixing of the liquids (and colors, to form green) will occur right away, with the accompanying evolution of heat and gas bubbles as hydrogen bonding and the differential sizes of the two types of molecules cause a reduction in volume. Even in the first setup, random molecular motion will eventually cause the two liquids to mix. If desired, students can see if the time this takes varies with the temperature of the fluids.

2. *Ethanol-Water Mixtures, Magic Money and Mileage: An STS Case Study*: The discrepant event in which a dollar bill soaked in a 50% water and 50% ethanol mixture will ignite and burn without harm (due to the high heat of vaporization of water relative to the bill's kindling temperature; see Internet Connections: About.com) can be used as an attention-grabbing introduction to ethanol-based biofuels. Ethanol-water and ethanol-gasoline mixtures (or gasohol) are currently being used as substitute automotive fuels in countries such as Brazil and the United States (see Internet Connections: Wikipedia). Students can research the STS tradeoffs involved in mixing ethanol with gasoline and devoting agricultural land and potential food products such as corn to the production of renewable biofuels to feed our internal combustion engines.

3. *"Waste Away" With Ecofoam Experiment, Demonstration, and STS Mini Case Study*: Petroleum-based, non-biodegradable, nonrenewable, water-insoluble Styrofoam (polystyrene) packing peanuts can be compared to starch-based, biodegradable, water-soluble ecofoam. Ecofoam packing peanuts will dissolve in a container of water without appreciably adding to the volume of the original water. This allows for a variety of "now you see it, now you don't" science magic tricks and art or construction projects that use water to glue together pieces of ecofoam. Ecofoam can also be detected with an iodine-based starch indicator. Students can explore and debate the tradeoffs by weighing the sources, relative benefits, and environmental burdens of these two alternative types of packing material (e.g., consider if boxes get wet during shipping, if they are stored in a location that could attract rats, or if large quantities of starch were added to natural waterways).

Measurements and Molecules Matter

They can also consider the importance of the eco-mantra "reduce, reuse, and recycle." See Internet Connections: Kansas City Public Television's e-Eats program.

Optional Teacher Demonstration: If a fume hood is available, a teacher (wearing safety goggles) can show how Styrofoam's 3D structure breaks down when it melts or dissolves in acetone (i.e., neither term is technically correct, as the acetone simply breaks some of the bonds that enable the polymer to maintain its shape). The small quantity of solid residue remaining (i.e., Styrofoam is mainly air) can be dried and thrown in the garbage, and the acetone can be reused many times. A video of this demonstration can be found in the Internet Connections: Steve Spangler Science.

4. *Repudiating the "Law" of Conservation of Volume*: Other dramatic examples of volume reduction include cases where large quantities of very water-soluble solids (e.g., table sugar or salt) can be added to water with little, if any, measurable increase in the volume of the solution relative to the starting volume of pure water. (*Note:* Total mass is conserved during the dissolution process.) Again, the respective intermolecular "holes" in water and the non-ionic solids and the interionic "holes" in salts allow for an increase in packing efficiency (i.e., this is in addition to the macroscopic spaces between the crystals of sugar or salt in the solid samples). The density of the solution formed is also greater than that of pure water.

5. *Laboratory Learning and Lesson Study*: Use the Internet Connections websites on lesson study and action research, as well as *America's Lab Report* (Singer, Hilton, and Schweingruber 2006), as starting points for a yearlong (or multiyear) lesson study or action research group focused on increasing the effectiveness of laboratory activities used in your school. See also Clark, Clough, and Berg (2000); Clough (2002); Clough and Clark (1994a, 1994b); Colburn (2004); Doran et al. (2002); Hofstein and Lunetta (2003); Lunetta, Hofstein, and Clough (2007); NSTA (2007c); and Volkman and Abell (2003) to further explore ways to redesign lab exercises to become inquiry-based explorations.

Internet Connections

- 3-D Molecular Designs (water model): *www.3dmoleculardesigns. com/news2.php#water*
- About.com: Chemistry: Burning Money Demonstration (uses a alcohol-water mixture): *http://chemistry.about.com/od/ demonstrationsexperiments/ss/burnmoney.htm*
- Alcohol + Water: A more complicated perspective on the nonadditivity of volume: *www.als.lbl.gov/als/science/sci_ archive/70methanolmix.html*
- *America's Lab Report: Investigations in High School Science*: *www.nap. edu/books/0309096715/html/R1.html*
- Cell Biology Animations (see: water): *www.johnkyrk.com*
- Center for Collaborative Action Research: *http://cadres.pepperdine. edu/ccar*
- Disney Educational Productions: *Bill Nye the Science Guy*: *Atoms* and *Measurement* ($29.99/26 min. DVD): *http://dep.disney.go.com*
- Doing Chemistry: Movies of chemical demonstrations: Mixing alcohol and water in a long tube: *http://chemmovies.unl.edu/ chemistry/dochem/DoChem070.html*
- Kansas City Public Television's e-Eats program: Ecofoam versus Styrofoam (PDF download): *www.kcpt.org/eats*
- National Academy of Sciences: National Research Council:

    Board on Science Education (*Framework for Science Education*) *http://www7.nationalacademies.org/bose/Standards_Framework_ Homepage.html*

    National Science Education Standards (1996): *www.nap.edu/ readingroom/books/nses/overview.html*

- Official M.C. Escher website: *www.mcescher.com*. The work titled *Relativity* makes a great prop when discussing the need for clear, consistent curriculum goals and instruction and assessment (CIA) that align with those goals. Both *Reptiles* and *Drawing Hands* can be used to represent the need for integrative, iterative cycles of CIA. *Bonds of Union* can represent teacher-student interactions.

Measurements and Molecules Matter

- Steve Spangler Science: Vanishing Styrofoam (description of experiment + video): *www.stevespanglerscience.com/experiment/00000046*
- Wikipedia: *http://en.wikipedia.org/wiki*. Search topics: action research, ehanol, ethanol fuel, and lesson study

Answers to Questions in Procedure, step #1

1. a. Yes, it matters how many decimal places are included because the different numbers indicate measuring equipment with different degrees of precision. Standard 100 ml graduated cylinders have 1 ml graduation marks, and volumes should be estimated to the nearest 0.1 ml.

   b. After mixing the two different liquids together, tiny gas bubbles are released or created (which might cause a minor volume reduction as they leave the liquid) and some heat is generated (which might cause a short-term volume expansion). Results for the combined volumes should near 96.0–97.0 ml. This is in contrast to the conservation of volume that occurs when a given liquid is added to itself (e.g., water + water or ethanol + ethanol).

   c. Excluding human errors such as misreading the volumes and spilling or splashing the liquids, a variety of factors might potentially contribute to a reduction in the final volume of the mixed liquids: some kind of chemical reaction that produced a new low-solubility gas that escapes and/or a physical change that generated heat that caused a previously dissolved gas to be expelled (in either case, there should be a corresponding reduction in mass that could be measured but is not found); rapid evaporation of one or both of the liquids (short of rapid boiling, this cannot account for the magnitude of volume reduction, and neither of the nonmixed liquids evaporate at a measurable rate in the time frame of the experiment); residual fluid left in the one "empty" graduated cylinder (the mass of this small volume can be checked with an accurate balance), and so on. None of these factors,

individually or collectively, can account for this order of magnitude in reduction of volume, and in fact heat alone would cause a slight, short-term expansion of the volume of the mixture. Something else must be going on. Someone may suggest that the two unknown liquids may have different densities (in fact, water has a density of 1.0 g/ml and ethanol 0.79 g/ml) and perhaps the denser liquid "squishes" the other less dense liquid into a smaller volume. This idea gets at a molecular view and the idea of different-size intermolecular "holes" or empty spaces. The next steps in the procedure will explore this hypothesis.

d. No question is asked in this step.

e. The marble-and-BB model clearly shows the idea that the two types of molecules have different sizes and different-size intermolecular holes that allow for some gains in packing efficiency when they are added together. If either marbles or BBs were added to themselves, the final volume would be additive (i.e., volume is conserved), as is true when any liquid is added to itself. Note that whether one adds the same kind of particles together or creates a mixture containing two different kinds of particles, the marble-and-BB model demonstrates the conservation of matter as consistent with doing the experiment with the actual liquids.

Major limitations of the marble-and-BB model include the following:

1. The "holes" are filled with air in the macroscopic model unlike truly "empty holes" that exist at the molecular level.
2. The model's static nature does not reflect the ever-changing location of the "holes" as related to the constant, dynamic motion (or kinetic energy) of the actual molecules in a liquid phase,
3. The model suggests that water and alcohol molecules are simply different size monatomic elements (such as helium versus neon) rather than more complex molecular chemical compounds.
4. The model does not reflect the existence or relative strength of intermolecular forces between like and unlike molecules.

Measurements and Molecules Matter

In the case of alcohol (marbles) and water (BBs), hydrogen bonding between the two different types of molecules (adhesion) is greater than the cohesive attraction (water-water and alcohol-alcohol), and these forces help contribute to the observed volume reduction. The more detailed space-filling molecular model kit shows the relative size, number, and arrangement of the constituent atoms in alcohol and water molecules, but the model kit also fails to represent either the hydrogen bonding or the dynamic molecular motions. (And, of course, atoms are not colored!) Neither of these two physical models demonstrates the generation of heat and the associated slight loss of dissolved gases (primarily nitrogen and oxygen driven out of solution by the heat) that are observed in the actual setup. Physical models, like analogies, always differ from the real systems they represent, but they provide both explanatory and predictive power to advance the scientific enterprise.

f. If the two different types of molecules repelled each other with enough force, then the final volume could be greater than the simple addition of their individual volumes, as the molecular "holes" would be larger. One example of this phenomenon is the addition of carbon disulfide to ethyl acetate. Chemical reactions between two different liquids that produced new chemical compounds could also conceivably lead to an increase in volume.

Safety Note

This teacher demonstration should not be done unless a fume hood is available and other safety precautions are followed.

Activity 16

Bottle Band Basics
A Pitch for Sound Science

Expected Outcome

Three identical-size glass soft drink bottles with varying amounts of water are used to make sounds of various pitches. Upon initial consideration, the pattern established when using the bottles as wind instruments seems to suggest a different rule than when they are used to produce different pitches as percussion instruments. In fact, the same underlying scientific principles explain the pitch patterns produced in all musical instruments.

Science Concepts

Sound is a form of energy produced by the vibration of matter where the pitch (or audio frequency) is, as a rough first approximation, inversely proportional to the amount of matter that is vibrating. Scientists are always looking for unifying theories, principles, and laws that bring together seemingly diverse observations about nature. Parsimony in explanatory frameworks is desired in science. More generally, this activity also emphasizes the idea of using more than only our sense of vision to find and explain patterns in nature.

Science Education Concepts

Sound, standards-based, science curriculum-instruction-assessment (CIA) puts "FUNomena first" to engage students' five senses, activate attention, catalyze cognitive processing, and teach students how to do science inquiry rather than simply receive science facts from an authority (i.e., the teacher or textbook). The activity serves as a *visual* and *auditory participatory analogy* for sound science where learners are challenged to learn how to use both teacher demonstrations and their own hands-on explorations to hoe the garden of their own minds to cultivate the seeds sown by teachers and by encounters with phenomena. Meaningful learning that is retained, retrievable, and transferable cannot be done to or for learners by the actions of teachers. Each student has to learn to "creatively play his own instrument."

Materials

- 3 identical-size glass soft drink bottles
- *Optional*: CD of soft instrumental music, CD of sound effects, an electronic tuner, and 3 identical wine glasses (the latter for an Extensions activity)

Bottle Band Basics

Points to Ponder

The mind, in short, works on the data it receives very much as a sculptor works on his block of stone. In a sense the statue stood there from eternity. But there were a thousand different ones beside it, and the sculptor alone is to thank for having extracted this one from the rest. Just so with the world of each of us, howsoever different our several views of it may be, all lay embedded in the primordial chaos of sensations, which gave the mere matter to the thought of all of us indifferently.

—William James, American psychologist and philosopher (1842–1910)

Equipped with his five senses, man explores the universe around him and calls the adventure Science.

—Edwin Powell Hubble, American astronomer (1889–1953)

We shall not cease from exploration and the end of all our exploring will be to arrive where we started and know the place for the first time.

—T. S. Eliot, British (U.S.-born) critic, dramatist, and poet (1888–1965), "*Little Gidding*," from *Four Quartets*

Procedure

Steps #3 and #4 can be done as a teacher demonstration and/or hands-on exploration.

1. If desired, begin the activity by playing some soft, relaxing instrumental music in the background as you initiate a brief conversation to activate and assess the learners' prior knowledge about sound. Ask: What is sound? What do you need to produce or generate sound? What do we need to detect sound? What are the evolutionary benefits of organisms being able to detect and differentiate various sounds? Are there sounds that humans can't

hear? And, for fun, if a tree falls in a forest and nobody is there, does it still make a sound?

Note: Learners are unlikely to offer all of the "answers" (provided on pp. 198–199), and the teacher should avoid the temptation to slide into the familiar teaching-as-telling mode or "words before wow and wonder." Rather, the teacher should simply try to gauge where the learners are with respect to their understanding of and interest in sound. If time permits, consider playing a quick game where you challenge the learners to name that sound from a sound effects CD.

2. Ask: How many different ways can we use glass soda bottles to make sound or music?

   After eliciting ideas, demonstrate the use of a bottle as a simple, one-pitch percussion instrument (unlike a wind instrument) by alternately tapping its side with a metal object such as a pen or pointer and blowing across the mouth of the bottle. Ask: How can we vary the volume (intensity or amplitude) of sound made by a given size soda bottle? How can we vary the pitch (frequency) to create music?

3. Predict: How will the pitch of a series of partially water-filled glass soda bottles vary with the amount of water the bottles contain? Will the relationship between the volume of water and the corresponding pitch differ if the bottles are struck (as a drum or percussion instrument) rather than blown across (as a wind instrument)? Why or why not? Briefly discuss their predictions and rationales without "sounding off" with the correct answer.

4. Observe: First, test the relationship for the bottle as a wind instrument and note the pattern. Construct a table to compare and contrast these results to those obtained when the same bottles are used as percussion instruments. (*Note:* The same instrument gives either the highest or lowest note depending on how it is played.) If desired, the bottles can be tuned to specific frequencies (by ear or with an electronic tuner) and musically talented teachers or learners can perform simple songs on a set of three to five bottles.

5. Explain: What underlying principle of sound enables instruments to be systematically tuned to specific pitches (or frequencies)? Is there a common underlying "rule" or "pitch pattern" that explains the behavior of all instruments? If so, what is it?

Debriefing

This is a simple, fun, and initially discrepant system that leads to the discovery of an underlying pattern that explains pitch variation across a variety of instruments. For other "sound science" activities, see Activities #6, #7, #10, and #13 in *Brain-Powered Science*.

When Working With Teachers

Use the quotes to generate discussions on the sensory basis for science and science learning and the importance of developing science understanding by scaffolding instruction from concrete FUNomena to the more abstract facts and principles that allow us to see familiar objects and phenomena in new ways. Sound science teaching is more about raising engaging questions to explore rather than giving answers to questions that were not raised by either phenomena or students. Interested teachers can explore the research on student preconceptions related to sound in chapters 13, "Air," and 18, "Sound," in Driver et al. (1994).

When Working With Students

Challenge students to work in teams to tune and play simple songs with their bottle bands (or do the Extension variations). The pitches and acoustical qualities of musical instruments are actually more complex than simply considering the *amount* of matter that is vibrating. Within a series of wind instruments such as the piccolo, flute, alto flute, and bass flute (or a series of progressively longer pipes on a panpipe), the fundamental, non-fingered note is inversely proportional to the *length* (not the mass) of the column of air. But higher pitches or harmonics can also be achieved by increasing the air pressure by blowing harder. Similarly, the pitches of the strings on a violin, viola, cello, and stand-up bass are inversely proportional to the length of the string (i.e., shorter strings have higher pitches). This can be demonstrated on a guitar by sliding down the fret board while plucking one string. However, stringed instruments can also achieve different pitches by increasing the string's tension (to increase its pitch) or increasing its mass with metallic wrapping (to decrease its pitch). Demonstrating a connection between music and science is another way to break down the artificial boundaries between science and art and help students see science as relevant to their lives and interests.

Extensions

1. *Resonance: Magic, Music, and Mayhem*: Another way to make music with a partially water-filled bottle without touching it is to hold a vibrating tuning fork over its mouth. A bottle whose wind instrument pitch matches the frequency of the tuning fork will "sing" in resonance with the vibrating fork. You may need to adjust the water in one of the bottles ahead of time to achieve this match depending on the frequency of the tuning fork. Alternatively, a glass or PVC tube partially submerged in a beaker of water can be raised or lowered to create the proper-length air column to match the frequency of a given tuning fork. Similarly, stringed acoustic instruments use the resonance of a sound board and chamber to amplify the sound generated by vibrating strings. Students can explore the concept of resonance as it applies to a variety of phenomena, from musical instruments to large-scale construction projects such as bridges (see Internet Connections: Tacoma Narrows Bridge Collapse).

2. *"Heavy Liquid" Music*: Challenge students to predict-observe-explain the effect of using equal volumes of liquids of different densities in identical glass bottles (e.g., ethanol = 0.76 g/ml; water = 1.0 g/ml; glycerin = 1.27 g/ml) used alternately as wind and percussion instruments. Will denser liquids make a difference in either or both types of instruments? Why? This activity could be done either as a teacher demonstration or with teams of students (wearing safety goggles) testing different variations.

3. *"Singing or Whining" Wine Glasses*: Ben Franklin invented the glass armonica, which consisted of a series of concentric glass bowls of increasing size mounted on a spindle (and suspended into a water trough so that they just barely touched) that rotated as a foot pedal was pumped. The player could play individual notes and chords by touching his or her fingers on the appropriate bowls. A series of identical glasses (e.g., wine) with varying amounts of water can serve as a model armonica. Each glass will emit a certain frequency when a wet finger is lightly rubbed around the wet rims. Before playing a series of glasses, you may wish to clean the oil off your index fingers by washing them in a little vinegar to maximize friction because the vibration is created

by the fingers alternately catching and sliding in quick succession. Ask: Do "singing" wine glasses behave more like a wind or percussion instrument, and why? Predict-observe-explain. (*Note:* The answer is that they behave like a percussion instrument [i.e., the more water in the glass, the lower the sound] because the entire glass + water + air system is set in vibration when friction is applied to the rim, which is the equivalent of striking.) Student teams can work together to tune a set of glasses and play specific songs (similar to handbell choirs) in playful competition with other student teams. See the Franklin Institute, Glass Armonica, and Wikipedia websites in Internet Connections.

4. *The Mathematics of Music* was first explored by the Greek philosopher Pythagoras of Samos (582–507 BC) and his followers. They blended a devotion to mathematics-music-mysticism-monaticism. See MacTutor History of Mathematics, University of Virginia Physics Department, and Wikipedia websites in Internet Connections, and chapter 9 in Joy Hakim's book *The Story of Science: Aristotle Leads the Way* (2004) to explore this fascinating mix of topics. Students' aptitudes and affective responses to music provide a motivational entry point to encourage interest in mathematics, the science of sound, and astronomy (i.e., "the music of the spheres"). Encourage voluntary student performances that creatively blend individual musical intelligence and Internet and empirical research on the science of sound. Also, free and for-purchase downloads of video clips and audio tracks of humorous science songs are available to enliven a wide variety of science topics (see Internet Connections: Sources of Science Songs).

5. *Aesop's Analogical Fable: The North Wind and the Sun (Teachers Only):* Consider the Bottle Band Basics activity as an example of the relative effectiveness of perplexing puzzles and engaged self-motivation (i.e., the gentle warmth of the Sun) to persuade, encourage, or inspire students to "take off the old coats" of their prior misconceptions versus instructional approaches that try to force student learning via intimidation and indoctrination (i.e., blowing hard like the wind). See the Internet Connections: Wikipedia.

6. *Human Noise: A Whale of an STS Problem*: Students can investigate the long-range communication network of whales and concerns that naval sonar research and the increased noise of international shipping are disrupting their "conversations" and possibly interfering with their eco-location of food and mating "songs."

Internet Connections

- Arbor Scientific's *Cool Stuff Newsletter*. See Sound & Waves listings (17 demonstrations): *www.arborsci.com/CoolStuff/Archives3.aspx*
- Disney Educational Productions: *Bill Nye the Science Guy: Sound* and *Science of Music* ($29.99/26 min. DVD): *http://dep.disney.go.com*
- Franklin Institute: Ben Franklin: Musician (listen and play the armonica): *www.fi.edu/franklin/musician/musician.html*
- Glass Armonica: Ben Franklin and the modern music of William Zeitler: *www.glassarmonica.com*
- How Stuff Works: Wine Glass "Singing" (video demonstration): *http://science.howstuffworks.com/question603.htm*
- HyperPhysics, Department of Physics and Astronomy, Georgia State University: Sound and Hearing: explanations, concept maps, and applications: *http://hyperphysics.phy-astr.gsu.edu/hbase/hframe.html*
- Learning and Teaching Scotland: Science of Sound Animations (8): *www.ltscotland.org.uk/5to14/resources/science/sound/index.asp*
- MacTutor History of Mathematics: Pythagoras of Samos: *www-history.mcs.st-and.ac.uk/Biographies/Pythagoras.html*
- Pacific Science Center: Music Physics and Instruments: *http://exhibits.pacsci.org/music*
- PhET Interactive Simulations: Sound: See sound waves and adjust frequency, volume, and location: *http://phet.colorado.edu/simulations/index.php?cat=Physics*
- Physics Classroom: Sound Waves and Music: *www.physicsclassroom.com/class/sound*
- Sources of Science Songs:

Dr. Chordate (various artists): *www.tranquility.net/~scimusic/notochordsproducts.html*

Greg Crowther's Science Song Music: *http://faculty.washington.edu/crowther/Misc/Songs*

Math and Science Song Information, Viewable Everywhere (MASSIVE) database: *www.science-groove.org/MASSIVE*

Mike Offutt Science CDs (for sale): *www.teachersource.com/Chemistry/ChemistryResources/Songs%20By%20Mike%20Offutt.aspx*

Science Song Links: *www.haverford.edu/physics/songs/links.html*

Songs for Teaching: Science Songs: *www.songsforteaching.com/sciencesongs.htm*

WIRED Science: Top 10 Scientific Music Videos: *www.wired.com/wiredscience/2009/07/sciencemusic*

- Tacoma Narrows Bridge Collapse:

    http://en.wikipedia.org/wiki/Tacoma_Narrows_Bridge

    www.youtube.com/watch?v=j-zczJXSxnw

    www.ketchum.org/bridgecollapse.html

- University of Iowa Physics & Astronomy Lecture Demonstrations (video clips): Acoustics: *http://faraday.physics.uiowa.edu* (demonstrations on air column instruments and more)

- University of New South Wales, Physics Department: Music Acoustics: *www.phys.unsw.edu.au/music*

- University of Toronto, Department of Physics: Physics Flash Animations: Sound: www.upscale.utoronto.ca/GeneralInterest/Harrison/Flash

- University of Virginia Physics Department: Mathematics of Music (a CBL) and Sound Activity Stations:

    http://galileo.phys.virginia.edu/outreach/8thGradeSOL/MusicMathFrm.htm

    http://galileo.phys.virginia.edu/outreach/8thGradeSOL/SoundStationsFrm.htm

> *http://galileo.phys.virginia.edu/outreach/8thGradeSOL/ WavePitchFrm.htm* (soda bottles)

- Whales Songs and Human Noise:

  http://en.wikipedia.org/wiki/Whale_song

  http://news.nationalgeographic.com/news/2002/06/0619_020619_ TVwhale.html

- Wikipedia: *http://en.wikipedia.org/wiki*. Search topics: glass harmonica, musical acoustics, Pythagoras of Samos, and the North Wind and the Sun

Answers to Questions in Procedure, steps #1–#5

1. Sound is a vibration of matter that can be heard by living organisms or detectible by instruments as a longitudinal or compression wave that travels through matter. In some cases, vibrations also may be sensed by directly touching the vibrating object or by feeling the vibrations through an intermediate fluid. Certain frequencies of vibration may even be visible in certain objects. Infrasounds and ultrasounds are those that are too low or high in pitch (i.e., frequencies below 20 cycles/second or 20 Hz, or those above 20,000 Hz) to be heard by humans, even though they may be heard by other organisms. Detecting sounds allows organisms to do a certain amount of remote sensing that enables them to make educated guesses (or informed inferences) about potential food, predators, mates, physical dangers, or other things that affect their short- or long-term survival (including those that are otherwise out of sight). Different organisms may experience dramatically different auditory realities even when they coexist in the same environment.

2. The volume of sound that is produced can be increased by striking the bottle with more force or by blowing harder across the bottle's mouth. The pitch can be changed by varying the relative amounts of air versus water in the bottle.

3. Many learners are likely to predict that pitch will be an inherent, intrinsic property of a given bottle (with a specific amount of

air and water) regardless of how it is played. Some learners will not have a clue and others (perhaps the musicians) will realize that one has to consider the material substance(s) that is (are) physically vibrating to generate the audible sound by vibrating surrounding air molecules.

4. Learners will observe the following, perhaps initially discrepant, pitch patterns:

Type of Instrument	Lowest Pitch	Highest Pitch
Wind: Air is blown across the opening of bottles.	Bottle with the most air (and least water)	Bottle with the least air (and most water)
Percussion: The bottles are physically struck.	Bottle with the least air (and most water)	Bottle with the most air (and least water)

5. Explain: The pattern of the empirical evidence from this activity suggests that the greater the mass of the matter that is vibrating, the lower the pitch. The actual physics of music is a bit more complicated than this simple experiment suggests. In wind instruments, it is the vibration of a column of air that generates the audible sound. Therefore, the bottle with the longest column of air (i.e., the least amount of water) makes the lowest pitch. If learners experimentally compare the sound generated by two equal-length, but different-diameter, air-filled test tubes, they will find that the wide and narrow test tubes have the same fundamental pitch. The critical factor affecting the pitch of wind instruments is actually the length, not the mass (or volume), of the air column. In contrast, a percussion instrument generates sound by vibrating the entire instrument and its contents. Therefore, the bottle with the most total mass (which is the one with the most water) has the lowest pitch. Some percussion instruments (e.g., a timpani) can also be tuned by increasing the tension in a membrane. The Internet Connections include sites where learners can probe more deeply into the physics of sound and music.

Metric Measurements, Magnitudes, and Mathematics
Connections Matter in Science

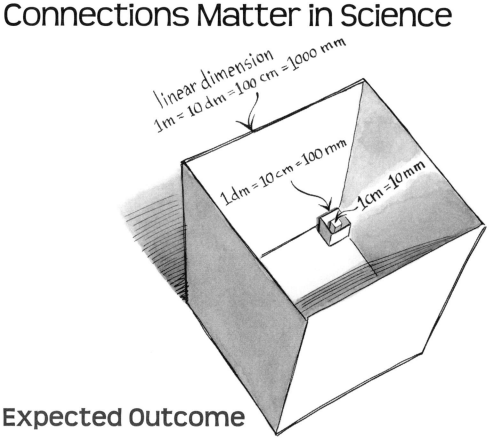

Expected Outcome

A cubic meter, assembled from 12 meter-long dowel rods, is used to develop an understanding of the relationships among metric units of length, volume, and mass. In combination with models of a cubic decimeter, cubic centimeter, and cubic millimeter, a cubic meter also serves as a concrete visual model for the idea of powers of ten and its real-world, cross-disciplinary applications.

Science Concepts

Given the SI units (*Système International d'Unités*, or the International System of Units) commonly used for metric volume measurements (e.g., 1,000 ml = 1 L), students may not realize that volume is actually always derived from the cube of a linear measurement (e.g., 1,000 cm^3 = 1 dm^3 = 1 L). Comparable English units of volume are problematic in that there are no easily remembered conversion factors within and between units such as ounces, pints, quarts, and gallons and the cubes of appropriately sized linear dimensions (e.g., in.^{3}). But, independent of units, learners of all ages also have difficulty conceptualizing very large numbers (e.g., mega = 10^6, giga = 10^9, tera = 10^{12}, very small linear dimensions (e.g., micro- = 10^{-6}, nano- = 10^{-9}, pico- = 10^{-12} meters), and very low concentrations (e.g., parts per million and parts per billion). The metric system, in addition to being simpler to use and to make interunit conversions, provides a concrete way to help students grasp the concept of powers of ten and the worlds of the very large and very small.

This model-building activity helps learners visualize SI's base-ten relationships; the connections between the meter, liter, and gram; the real-world relevance of mathematics to science; and the somewhat abstract concept of powers of ten that is so central to science and society. As such, it also serves as a means to highlight and confront innumeracy or mathematical illiteracy. Additionally, the activity serves as a physical analogical model for scientific entities that exist as related "boxes within boxes" of nested realities that generate new, emergent properties at each higher level of system organization (e.g., subatomic particles → atoms → compounds → mixtures → living cells → organisms → ecosystems → biosphere; see Morowitz 2002).

Science Education Concepts

The AAAS Project 2061 *Benchmarks for Science Literacy* (AAAS 1993, chapters 2 and 9) and the *National Science Education Standards* (NRC 1996; p. 104) call for contextualizing science concepts in the big picture frame of, respectively, four common themes or five unifying concepts and processes. Similarly, *A Framework for Science Education, Preliminary Public Draft* (NRC 2010, chapter 4) cites seven cross-cutting science concepts. Regardless of the specific number and terms used,

Metric Measurements, Magnitudes, and Mathematics

explicit attention to the "forest" helps students understand and meaningfully connect the numerous smaller "trees" of science. *Benchmarks* identifies *systems* (i.e., parts to whole relationships), *models* (e.g., physical, conceptual, and mathematical), *constancy and change,* and *scale* (i.e., powers of ten) as four common themes that should pervade all science courses (AAAS 1993, pp. 261–279). It also calls for the intentional integration of mathematics in science (*Benchmarks,* chapters 2, 9, and 11) as does the *National Science Education Standards* (NRC 1996, pp. 116–117, 148, 175–176, 214–218), while the *Framework*'s *concepts* are based on an integrated view of science, technology, engineering, and mathematics (STEM). This activity addresses the four common themes, mathematics, and real-world STEM applications in the context of measurement.

This activity also serves as a *visual participatory analogy* to stress that effective science education is not racing students through a bunch of unrelated, disconnected "boxes" to be mindlessly memorized and soon forgotten. An effective science curriculum-instruction-assessment (CIA) system is based on a view of science and learning as constructive processes that make sense and generate synergistic outcomes where the whole is greater than the sum of the parts. Memory retention, recall, and transfer of learning to new contexts are enhanced when learners are challenged to reconstruct their preinstructional conceptual networks to better reflect the known, currently accepted, interrelated scientific concepts and theories. If desired, this activity can be paired with an introduction to *concept mapping*, which is a type of graphic organizing tool that makes visually explicit both the nested-reality nature of scientific concepts and learners' conceptual networks or mental models about such concepts (Good, Novak, and Wandersee 1990; Mintzes, Wandersee, and Novak 1998, 2000). See the Internet Connections: Institute for Human and Machine Cognition and Wikipedia.

Materials

- 1 (or more) cubic meter models can be constructed (for about $7 each) in the following manner: Cut 12, 1/4 in. × 46 in. dowel rods down to 100 cm (or 39.37 in.) in length. Cut 8 connecting corner piece cubes to be approximately 3–4 cm^3 wood cubes from a piece of 2 in. × 2 in. lumber. Drill holes all the way through the centers

of 3 of the 6 faces (i.e., 2 adjacent sides and the top of each cube), through which the dowel rods can be firmly inserted. The *outside* linear dimensions of the assembled 3D cubic meter (i.e., outside edge of one corner to the outside edge of another corner piece) should be 1 m.
- Learners should also have a 15 or 30 cm (6 or 12 in.) ruler and a hand lens or handheld microscope (see Activity #8 for sources) for a close-up view of table salt crystals (approximately 1 mm^3 cubes). Cubic decimeters (or L) and cubic centimeters (or ml) can be constructed with poster board or purchased commercially from one of the sources listed.
- Individual crystals of table salt will serve as approximations of cubic millimeters.
  - Suppliers of Metric Measuring Equipment
    - *EAI Education*: liter cube with lid (clear plastic): #530195/$5.95: *www.eaieducation.com/530195.html*
    - *ETA Cuisenaire*: #IN4603: clear liter volume cube ($7.95): *www.etacuisenaire.com*
  - *Flinn Scientific*, Inc.: *www.flinnsci.com*; 800-452-1261
    - Liter cube set (AP4787/$17.75): 1 L cube + 9, 100 cm^3 flats; 9, 10 cm^3 rods, and 10, 1 cm^3
    - Becker Bottle "One in a Million" (AP4559/$33.95): 3 L plastic bottle with one million tiny, variable-colored spheres to visually demonstrate 1 ppm, 10 ppm, 100 ppm, and so on. Alternatively, one could make a homemade version by "counting" rice grains by weighing them and inserting the appropriate number of colored rice grains.
  - *Nasco Math and Science*: *www.eNasco.com*; 800-558-9595. Variety of metric products:
    - Interlocking gram centimeter cubes (#TA02278M; $20.50): 1,000, 1 cm^3 cubes that have a mass of 1 g each and that can be combined in various ways.
    - Liter cube with lid (#SB28838M; $6.95)

Metric Measurements, Magnitudes, and Mathematics

> ## Points to Ponder
>
> *When we try to pick out anything by itself, we find it hitched to everything in the universe.*
>
> —John Muir (1838–1914), American conservationist
>
> *In training a child to activity of thought, above all things we must beware of what I call 'inert ideas'—that is to say, ideas that are merely received into the mind without being utilized, or tested, or thrown into fresh combinations ... Education with inert ideas is not only useless: it is above all things, harmful ... You may not divide the seamless coat of learning. What education has to impart is an intimate sense for the power of ideas, for the beauty of ideas, and for the structure of ideas, together with a particular body of knowledge which has peculiar reference to the life of the being possessing it.*
>
> —Alfred North Whitehead, English mathematician and philosopher (1861–1947) in his 1916 speech "The Aims of Education" (see Internet Connections)
>
> *Eventually, everything connects.*
>
> —Charles Eames, co-director of the classic film *Powers of Ten: About the Relative Size of Things in the Universe* (1977)

Procedure

When Working With Teachers or Students

Depending on the learners' prior experiences with the metric system, the initial guided-discovery questions in this section may be covered as a fairly quick review (i.e., teachers). For learners with less familiarity with the metric system (i.e., middle school students), more extended hands-on explorations would be needed to cover steps #1 and #2 before proceeding to metric units of volume (step #3), large numbers (step #4), and small concentrations (step #5). *Note:* Even

when working with teachers, doing justice to the last two steps (#4 and #5)—much less the numerous Extension activities and internet sites provided—will require more time than most of the other activities in this book. Scale is a really big idea in science that is typically given too little curricular attention and instructional time.

1. *Looking About With Linear Units*: What is the basic unit of length in the SI or metric system? Distribute meter sticks (and/or unmarked, 1 m long dowel rods) and smaller metric rulers for the learners to examine. What common objects around this room or school could be easily measured in meter-length multiples? What metric units would be used for measuring the linear dimensions (i.e., length, width, and height) of smaller objects such as a desktop, a book, paper clips, or the thickness of a staple? How do these units relate to the meter? How are these interunit relationships different from the equivalent comparisons of large and small linear units in the English system?

2. *Sizing It Up With Surface Area Units*: What unit would be used to determine how much carpeting would be needed to cover the floor of the classroom? Recruit several volunteers to help construct two (or more) separate square meters using four dowel rods and four corner pieces per square meter. If time permits, quickly make the linear measurements and calculate the surface area of the floor. Point out that the appropriate unit is a square meter even if the classroom is rectangular in shape. Challenge the learners to look around the classroom for other objects whose surface area is some multiple of a square meter.

    Given that a square meter is a rather large unit, what metric unit might be more appropriate for measuring the surface area of a smaller object, such as the amount of material needed to cover a textbook or laminate an ID card? Contrast the square meters that were assembled with both a 1 dm^2 (10 cm × 10 cm square) and a 1 cm^2 that learners can draw at their seats. What are the relationships between a square centimeter, square decimeter, and square meter? What are some objects around the classroom or in everyday life whose surface areas could be easily measured with the units of square centimeters or square decimeters versus square meters?

NATIONAL SCIENCE TEACHERS ASSOCIATION

Metric Measurements, Magnitudes, and Mathematics

Do objects have to be square or rectangular to measure their surface areas? Use an overhead projector or document camera to project an image of a circle overlaid on two different pieces of graph paper with relatively larger and much smaller square units to establish that surface area is always measured as a square of a linear dimension. If desired, time could be spent on measuring the linear dimensions and calculating the surface area of various size circles ($A = \pi r^2$), cubes ($A = 6 s^2$), and cylindrical cans ($2 \pi r^2$ for the two end caps + length × width of side wall). Determining the minimum amount of wrapping paper that would be needed to cover gifts is a practical application of this process.

3. *Verifying With Volume Units*: What units would we use if we wanted to determine the maximum amount of storage space in our classroom's closet or storeroom? How can we use the standard unit for linear measurement (the meter) to derive base units and build models for volume? Use four remaining 1 m rods to complete the construction of a cubic meter by connecting the two previously assembled square meters in the vertical dimension. Several volunteers can easily fit inside the cubic meter box. Use this large-scale physical model to help solidify the idea of volume as a cubic measure. What are some other school-related objects whose volumes could be expressed in cubic meters?

What is the problem with a cubic meter as a unit of volume? What is commonly considered the metric base unit of volume measurement, and what is its relationship to a meter and a cubic meter? How many cubic decimeters (or liters) would fit inside a cubic meter? What are some common objects from the classroom or everyday life whose volumes would be easily represented in multiples of one liter? One liter (or cubic decimeter) represents what fraction (or part) of a cubic meter? What unit do we commonly use to measure the volumes of objects that are significantly less than a liter, and how does this unit relate to a liter? Use physical models to visually demonstrate the answers to these questions, and take time to make sure that the learners can follow the mathematical relationships. *Note:* Sugar cubes are just a bit too large to properly represent a cubic centimeter, but commercially available plastic cubic centimeters (with a mass of 1 g each) are not very expensive.

If desired, you may also wish to define and demonstrate that 1 L of water has a mass of 1 kg (and because 1 kg = 1,000 g and 1 L = 1,000 ml = 1000 cm^3, 1 ml or 1 cm^3 of water has a mass of 1 g). This fact makes it possible to easily convert between the mass and volume of (pure) water without the need for a calculator or divisions. If time permits, learners can mass various quantities of water to determine the density of water (1 g/cm^3, which is the slope of a graph of mass versus volume).

4. *Making Millions (and Building Billions) With Metric Units*: How can we use these physical models of cubic meters, decimeters, and centimeters to visually represent large numbers such as millions (10^6) and billions (10^9)? How many cubic centimeters would fit into a cubic decimeter (or liter) versus a cubic meter? How might the smaller unit of cubic millimeters help us visualize a billion? Can you think of any common material whose volume is about 1 cubic millimeter?

After the learners have considered the mathematics of these questions, provide them with a hand lens and metric ruler to examine table salt (i.e., it consists of tiny cubes of sodium chloride that are each just slightly smaller than a cubic millimeter or 1 mm^3). Ask the learners to visualize a cubic meter box being filled with these tiny crystals of table salt stacked perfectly on top of and beside each other in neat, closely packed rows and columns. A cubic millimeter is just about the smallest volume we can measure with the naked eye and visualize being "added up" to equal a billion—giving new meaning to the phrase "Take it with a grain of salt."

5. *Concentrations Count: A Little May Mean a Lot*: Why should we care about minute concentrations of substances that are measured in parts per million (ppm) and parts per billion (ppb)? In what real-world contexts are these units used? See the Internet Connections for ppb and Ozone Model activity that includes visuals that compare mm^3 → cm^3 → dm^3 → m^3.

Metric Measurements, Magnitudes, and Mathematics

Debriefing

When Working With Teachers

Emphasize the synergistic use of physical and mathematical models to help students visualize and understand science concepts. Use the Points to Ponder quotes and physical models (as a visual participatory analogy) for how, in science, successively larger ideas provide the bigger picture context or "forest" that contextualizes and helps make sense of the smaller "trees" (and vice versa). Since the time of the Greek natural philosophers, scientists have attempted to discover the relational order or unity (or cosmos) that underlies the seemingly endless diversity (or chaos) of different natural phenomena and events. Big ideas (core theories) and common themes provide the unifying context that enables us to almost holographically "see a world in a grain of sand" (e.g., atoms produced in the nuclear furnaces of "dying" stars). The idea of scale is an especially critical theme that cuts across all sciences. Teachers should be very critical of the numerous textbook visuals that grossly misrepresent the relative scale of objects as small as solar-system-type atomic models, cutaway sketches of cells, and photomicrographs with unmarked scales to objects as large as the Earth-Moon system (O'Brien and Seager 2000), our solar system, and beyond. Off-scale representations should always be noted as such to avoid misrepresenting the truly awesome nature of reality. Concrete models and visual analogies should be used to correct these textbook misrepresentations or visual lies.

Discuss the power of connected, meaningful, real-world learning versus "inert ideas." Ask the teachers to review how the *Benchmarks for Science Literacy* (AAAS 1993), the *National Science Education Standards* (NRC 1996), and ongoing refinement of standards are challenging conventional teaching practices with respect to big picture ideas and themes. Develop and share other models for the idea of millions (or billions; e.g., a 40-dot-wide by 50-dot-long matrix placed on an 8.5 in. × 11 in. sheet of paper contains 2,000 dots; 500 of these pages contain 1 million dots. Calculate what portion of the classroom walls or floor would be covered by these 500 sheets of paper.). Also consider engaging in collaborative study groups with mathematics teachers using such books as John Allen Paulos's *Innumeracy:*

Mathematical Illiteracy and Its Consequences, A Mathematician Reads the Newspaper, and *Beyond Numeracy* or John Barrow's *Pi in the Sky: Counting, Thinking, and Being.*

When Working With Students

Challenge students to estimate and then measure the lengths, surface areas, and volumes of common objects. If desired, students can calculate (via the Pythagorean theorem of $a^2 + b^2 = c^2$ or $c = 1.4142$ m) and then measure the length of the cubic meter's diagonal with a string. The Extension activities, in combination with traditional hands-on explorations (e.g., measuring water's density versus that of other liquids and testing how $1g/cm^3$ blocks behave in different liquids), can help students construct the concept of mass relative to volume (see Activity #3 in this book for other density activities). Videos such as *The Powers of Ten*, computer animations, and Google Earth satellite images to view local school sites (see Internet Connections) help students visualize the range of scales of magnitudes in the universe. Scale is a major grades K–16 theme for achieving scientific literacy and promoting an informed sense of "wow and wonder" about our universe. Also, concept mapping is a research-validated instructional strategy to show students both how scientific ideas relate as "boxes within boxes" of nested concepts and theories and how students can critically examine how their thinking matches up to modern scientific conceptions (see the Internet Connections: Institute for Human and Machine Cognition).

Extensions

1. *Scales of Science: Magnitudes Matter.* Each science discipline has its own "far out" numbers that pose visual and conceptual challenges (see also Activity #2 in this book). Students can be challenged to use the cubic meter model or develop other relative scale models to represent various powers-of-ten comparisons, such as the following:

    Biology: Approximately 30×10^6 known species that range from nanometer-size viruses (20–250×10^{-9} m) to micrometer-size bacteria (the smallest, the mycoplasmas, are 0.1–1.0×10^{-6} m in diameter, while typical prokaryotic cells are 1–10×10^{-6} m)

Metric Measurements, Magnitudes, and Mathematics

to typical eukaryotic cells (10–100 × 10^{-6} m) up to the 31 m long blue whale, the largest macroscopic organism on Earth. DNA, RNA, proteins, and other complex biomolecules that are the foundation for all life exist at the nanometer-size scale. Research and development in nanoscience and engineering are increasingly tapping the wisdom of billions of years of evolution-based research and development and using biological systems as models and manufacturing machines (see the books listed under Nanoscale at NSTA Press on p. 215).

Challenge learners to assess the plausibility of this "far out" claim: "Humans contain more bacteria cells in and on their bodies than human cells!" Logical argument: V_{sphere} = 4/3 πr^3 (for spherical cells) or $V_{cylinder}$ = πr^2 × height (for rod-shaped cells). In all cases, volume is a cubic measurement, so factors of 10–100 differences in linear dimensions translate to factors of 1,000 → 1,000,000 in volume differences between the two types of cells. The cubic centimeter to cubic decimeter and cubic centimeter to cubic meter models can be used to help students visualize these comparative sizes. The thousand-fold drop in linear dimensions from the micrometer world of bacteria (1 μm = 10^{-6} m) to the nanometer world of viruses (1 nm = 10^{-9} m) is even more dramatic in terms of relative volumes (and requires the cubic millimeter to cubic meter or part per billion comparison). Images in many textbooks do not list magnifications on drawings or photomicrographs, nor do they represent the two kinds of cells and viruses on the same scale. As a result, students leave biology classes with misconceptions about the relative 3-D sizes of microscopic entities and only a vague idea of the challenges faced by germ theory pioneers and current-day medical biology researchers.

On the macroscopic scale, understanding human population dynamics (e.g., U.S. population = 310,838,535 = 4.5% of the world's population of 6,885,443,201), resource consumption, and pollution production also require powers-of-ten thinking. (See Internet Connectioins: U.S. Census Bureau for the most up-to-date numbers and statistics.)

Chemistry: A mole of any element or compound contains Avogadro's number (6.02×10^{23}) of particles. A mole of any gas at STP (0°C and 1 atm of pressure) occupies the same volume (22.4 L, or a box that is 28.189 cm per edge; a little more than 44.64 of these boxes would fit into the cubic meter box). This large number of particles points to the sub-nanometer-size world of atoms and molecules, while the relatively low density of gases points to the large intermolecular distances in gases.

Get students thinking about the large intra-atomic spaces by challenging them to assess the plausibility of this "far out" claim: "Most of the atom is empty space!" Logical argument: Atomic radii: picometers = 10^{-12} m. Range: 25 pm (H) – 260 pm (Cs). Nuclear radii: femtometer = 10^{-15} m. Range: 0.8×10^{-15} m (for a proton in light H) → 7.5×10^{-15} m (for the heaviest atoms, such as U). Given that $V_{sphere} = 4/3\ \pi r^3$, the ratio of the volume of an atom to the volume of a nucleus (using H as an example) is 3×10^{13} to 1! A ratio of one billion (10^{12}) to one can be represented by a cubic millimeter (= one crystal of table salt) in the middle of a cubic meter. Therefore, 10^{13} to one ratio would equal one salt crystal in the middle of 10 cubic meters (or a cube whose side was 2.15 meters long). Atomic nuclei are incredibly small, dense objects in the center of atoms that are mainly empty space.

Geology: The Earth is 4.6×10^9 years old with geological periods measured in millions of years back (mya) from the present era (see Internet Connections: Geological Society of America and the University of California Museum of Paleontology). Textbooks typically depict these periods of "deep time" and the associated evolution of life with visuals that have cuts or breaks that make them off-scale so that they can fit on a page. Metric measurements, models, and visual analogies can be used to help make these large numbers more concrete. For example, 1 km (0.62 mi.) contains 10^6, or one million, millimeters; therefore 1,000 km (or 620 mi.) = 10^9 mm. Students can be asked to visualize another city that is approximately this distance from their school and then think of a 1 mm segment as one year in a billion of geological time. The mm³ to

NATIONAL SCIENCE TEACHERS ASSOCIATION

Metric Measurements, Magnitudes, and Mathematics

m^3 model can also be used to visualize the idea of a billion. Alternatively, challenge learners to construct a scale model of geological time and the evolution of living species with a roll of toilet paper. Note the early emergence of prokaryotic cells and the late, last-minute arrival of humans (see O'Brien 2000).

Physics: Concepts that require an understanding of powers of ten include 3.0×10^5 km/s (speed of light in a vacuum), quantities of energy and matter, and the relative sizes and distances in astrophysics and astronomy versus quantum mechanics (see Internet Connections: *Powers of Ten*–type films and animations). Also, the wide range of frequencies (10^{24} MHz gamma rays to 10^2 MHz radio waves) and inversely proportional wavelengths (10^{-16} m to 10^6 m) across the various forms of electromagnetic radiation makes it impossible to represent the electromagnetic (EM) spectrum on a linear scale in a textbook. So, as with geological and evolutionary time, powers-of-ten scales and cutaway and truncated-type visuals are used. See Internet Connections: HyperPhysics.

2. *Metric Math and Merriment*: How much would a cubic meter of a typical type of Chinese soup "weigh"? Or, stated in a more scientifically correct manner, what is its mass? *Hint:* What is the main ingredient of many common types of soup? What is the relationship between the metric units for mass (gram) and volume (milliliter or cubic centimeter)? See Answers to Questions in Procedure, pages 218–220.

3. *Metric Math Matters to the Environment*: An environmentalist urges consumers to buy powdered laundry detergents rather than liquid ones. Assuming no significant differences in the relative cleansing power, cost, or environmental impact of the active ingredients, why does this make good sense in terms of both consumer economics and environmental ethics? See Answers to Questions in Procedure, page 219.

4. *Metric Math and the Biological Scale Effects*: The surface area (SA) to volume (V) ratios of the metric cubes used in this activity decrease by a factor of ten for each tenfold increase in linear dimension (1 mm^3 → 1 cm^3 → 1 dm^3 → 1 m^3). The mathematics

of SA-to-V ratios helps explain, for example, (a) how the addition of membrane-bound organelles allows eukaryotic cells to be so much larger than prokaryotes; (b) the "need" for multicellular organization (versus macro-sized cells); and (c) heat gain or loss and nutrition challenges faced by different-size animals (e.g., a shrew versus an elephant).

5. *Metric Math and Pollution Solutions Hands-On Exploration*: Parts per million (ppm) can also be visualized by doing a serial dilution of a 10% (ten parts per hundred) solution of food coloring. If nine drops (or any other standard volume measure) of water are added to one drop of liquid food coloring, a 1% solution (one part per hundred) is obtained; if nine drops of water are added to one drop of this 1% solution, a 0.1% solution (one part per thousand) is obtained. This process can continue until a 1 ppm solution is created. The color should no longer be visible at this concentration. If a colored mouthwash is used instead of food coloring, students may be able to detect the odor of the diluted mouthwash even after they can no longer see the color. The safety levels for many environmental pollutants are regulated at ppm concentrations, levels at which we probably would not be able to "see" the chemicals in water or air even if they were colored compounds. Serial dilutions also raise the pros and cons of the idea that dilution is the solution to pollution. See SEPUP's Investigating Wastewater and the American Society for Microbiology sites in Internet Connections.

6. *Metric Model for Molecular Level Phase Change*: Most textbook drawings (and even some molecular animations) misrepresent the changes in volume and density that occur when a liquid becomes a gas under constant atmospheric pressure. *Note:* A picture may be worth a thousand words, but pictures can lie. For water and most other substances, this phase change results in roughly a thousand-fold increase in volume (and decrease in density). This effect can be visualized by comparing 1 cm^3 to 1 dm^3 or 1 dm^3 (or 1 L) to 1 m^3.

7. *Metric Numeracy and ELA Literacy*: A number of children's books have tackled the difficult task of helping students grasp the powers-of-ten concept that underlies very big numbers by using

Metric Measurements, Magnitudes, and Mathematics

visual mathematical analogies and concrete examples that appeal to learners of all ages.

- Banyai, I. 1995. *Zoom.* New York: Viking/Penguin Books.
- Birch, D., and D. Grebu (illus.). 1988. *The king's chessboard.* New York: Penguin Books.
- Packard, E., and S. Murdocca (illus.). 2000. *Big numbers and pictures that show just how big they are!* Brookfield, CT: Millbrook Press.
- Schwartz, D. M., and S. Kellogg (illus.). 1985. *How much is a million?* New York: Lothrop, Lee & Shepard Books/HarperCollins.

8. *Mars Metric Mishap Conversion Calamity*: In September 1999, NASA lost the $125 million Mars Climate Orbiter. "The 'root cause' of the loss of the spacecraft was the failed translation of English units into metric units in a segment of ground-based, navigation-related mission software." This story presents a real-world argument for the United States making a complete conversion to the internationally accepted metric system. See Internet Connections (Mars Metric Mishap Conversion Calamity JPL/NASA website).

9. Nanoscale at NSTA Press: *http://www3.nsta.org/presshome*

    - Jones, M. G., M. R. Falvo, A. R. Taylor, and B. P. Broadwell. 2007. *Nanoscale science activities for grades 6–12.*
    - Jones, M. G., A. R. Taylor, and M. R. Falvo. 2009. *Extreme science: From nano to galactic.*
    - Stevens, S. Y., L. M. Sutherland, and J. S. Krajcik. 2009. *The big ideas of nanoscale science and engineering*: A guidebook for secondary teachers. Unlike the two previous books, this one does not contain activities for use with students but, rather, is intended to help teachers with relevant science content knowledge.

    All three books show how nanoscience concepts bridge the various disciplines of science to help us form an integrated view of our world.

Internet Connections

- *Aims of Education*: 1916 Mathematical Association presidential address and 1929 book by Alfred North Whitehead: *www.ditext.com/whitehead/aims.html*

- American Society for Microbiology: Classroom Activities: *Is It Clean or Just Unseen? Dirty Water and the Naked Eye* (serial dilution + bacterial cell count): *www.asm.org/index.php/education/classroom-activities.html*

- Arbor Scientific's *Cool Stuff Newsletter*. See Measurement: Molecular Monolayers (uses a BB/oleic acid molecule analogy to calculate the size of a molecule): *www.arborsci.com/CoolStuff/Archives3.aspx*

- *Cosmic View: The Universe in 40 Steps* (1957 book by Kees Boeke): *http://nedwww.ipac.caltech.edu/level5/Boeke/frames.html*

- Cubic Meter Model Instructions: *http://mynasadata.larc.nasa.gov/docs/Metric_Cube.pdf*

- Geological Society of America: Geological Time Scale: *www.geosociety.org/science/timescale*

- Google Earth: *http://earth.google.com* (free software download)

- How Stuff Works: Nanotechnology: *http://science.howstuffworks.com/nanotechnology.htm*

- HyperPhysics: Electromagnetic Spectrum: *http://hyperphysics.phy-astr.gsu.edu/hbase/ems1.html#c1*

- Institute for Human and Machine Cognition, *free CMap Tool*: *http://cmap.ihmc.us*

- Java Applets on Mathematics: Unit Conversion, Pythagorean Theorem I and II, and the Circumference and Area of the Circle: *www.walter-fendt.de/m14e*

- *The Known Universe* (American Museum of Natural History powers-of-ten–type film travels from the Himalayan Mountains out 13.7 billion light years and back to Earth in 6:31 min): *http://apod.nasa.gov/apod/ap100120.html*

- m^3 Model: *www.phys.unt.edu/~klittler/demo_room/mech_demos/1a10_75.html*

Metric Measurements, Magnitudes, and Mathematics

- Mars Metric Mishap Conversion Calamity:
    http://mars.jpl.nasa.gov/msp98/orbiter
    www.cnn.com/TECH/space/9909/30/mars.metric.02
- Metric Conversions: *www.metric-conversions.org/volume/cubic-meter-conversion.htm*
- *Microcosm*, CERN's Interactive Science Centre: Powers of Ten: steps through 10^{-15}–10^{26} meters views: *http://microcosm.web.cern.ch/Microcosm/P10/english/welcome.html*
- NanoScale Science Education Research Group at North Carolina State University: *http://ced.ncsu.edu/nanoscale*
- The NanoTechnology Group, Consortium for Global Education (variety of resources): *www.tntg.org/documents/home.html*
- National Council of Teachers of Mathematics: National Metric Week (10/10/10): *www.nctm.org/news/content.aspx?id=10248*
- National Nanotechnology Infrastructure Network: Size and (relative) Scale Curriculum Unit: *www.nnin.org/nnin_k12sizeandscale.html*
- NSF's National Nanotechnology Initiative: *www.nsf.gov/crssprgm/nano/education/kids.jsp*
- PhET Interactive Simulations: Estimation (game) for length, surface area, and volume: *http://phet.colorado.edu/simulations/sims.php?sim=Estimation*
- ppb and Ozone Model activity: *www.globe.gov/tctg/modelppb.pdf?sectionId=27*
- *Powers of Ten* (1977 Charles and Ray Eames film): *http://powersof10.com*
- *Secret Worlds, the Universe Within*: free, interactive Java powers-of-ten-type video: *http://micro.magnet.fsu.edu/primer/java/scienceopticsu/powersof10*
- SEPUP's *Investigating Wastewater: Solutions and Pollution* module: *wwww.sepup.com/catalog.php?item=SP-2*
- University of California Museum of Paleontology: Web Geological Time Machine: *www.ucmp.berkeley.edu/help/timeform.html*

- U.S. Census Bureau: (the ideas of millions and billions of human beings):

    U.S. and World Population Clocks: *www.census.gov/main/www/popclock.html*

    International Data Base: *www.census.gov/ipc/www/idb*

- U.S. Treasury Department, National Public Debt Report (trillions $): *www.treasurydirect.gov/govt/reports/pd/pd.htm*
- Wikipedia: *http://en.wikipedia.org/wiki*. Search topics: concept map, *Cosmic Voyage* (1996 IMAX film; depicts 42 orders of magnitude; similar to *Powers of Ten*), cubic meter, density of universe is less than 1 atom/m^3, metric system, nanotechnology, orders of magnitude (length), parts-per notation, and *Powers of Ten* film

Answers to Questions in Procedure, steps #1–#5

1. The meter is the metric base unit of length. The linear dimensions of a school bus, a classroom, a teacher's desk, a door, and students' heights could all be measured in terms of multiples of a meter. Smaller objects such as the students' desktops, books, and pencils could be measured in decimeters (1 dm = 0.1 m), and chalk and paper clips in centimeters (1 cm = 0.1 dm = 0.01 m). Even smaller objects such as a staple would be measured in millimeters (1 mm = 0.1 cm = 0.01 dm = 0.001 m). The fact that metric units are based on multiples of ten makes for easy conversions between different size units (e.g., 10 mm = 1 cm; 10 cm = 1 dm; 10 dm = 1 m) in contrast to the English system (e.g., 12 in. = 1 ft.; 3 ft. = 1 yd.; and 1,760 yd. = 1 mi.).

2. Surface area is measured in terms of squared linear units. The constructed square meter models can be held against the surface of the classroom's floor; black, white, or smart board; projection screen; door; and teacher's desk to see that the square meter (m^2) is the appropriate unit size. If desired, students can be asked to make quick estimates or actual measurements. The surface area of smaller objects such as textbooks and ID cards

could be measured in terms of square decimeters (dm^2) or square centimeters (cm^2) where the relationships are 1 m^2 = 10 dm × 10 dm (or 100 dm^2) = 100 cm × 100 cm (or 10,000 cm^2). The surface area of any object, regardless of its shape, is measured in linear units squared. If progressively smaller-square graph paper is superimposed on a projected image of a circle, learners should be able to visualize the validity of this statement as a process of successive approximation. The exercise of measuring the dimensions and calculating the surface area of cubes and cans (or cylinders with two lids) can be used to extend this idea to three-dimensional objects.

3. The volume of a closet could be measured in terms of cubic meter (m^3), as could the volume of a school bus and a swimming pool. A cubic meter is too large a unit to represent the volumes of many common solid objects or quantities of liquids that we can carry or consume. One liter (1 L) = 0.1 m^3 = 1 dm^3 is commonly considered the base unit for metric volume. Ten 1 dm^3 boxes could fit along each of three dimensions of the cubic meter, so 10 dm × 10 dm × 10 dm = 1,000 cubic decimeter boxes would fill the cubic meter. A mug of coffee, soft drink can, container of milk, and container of gasoline are all examples of objects whose volumes would be measured in liters. If one is discussing concentrations, a 1 L or 1 dm^3 box is one part per thousand of a cubic meter. Smaller volumes are measured in cubic centimeters (1,000 cm^3 = 1,000 ml = 1 L = 1 dm^3).

4. One hundred (100) cm^3 boxes could fit along each of the three dimensions of a cubic meter, so 100 cm × 100 cm × 100 cm = 10^6 cubic centimeters would fit into a cubic meter. Therefore, 1 cm^3 is one part per million (ppm) of a cubic meter. By a similar calculation, 1 cm^3 is one part per thousand of a cubic decimeter (or liter). Each 1 m side of the cubic meter is 1,000 mm in length, therefore 1,000 mm × 1,000 mm × 1,000 mm = 10^9 mm^3. One cubic millimeter or 1 mm^3 = 1 part per billion (ppb) of a cubic meter. *Note:* A standard 26 oz. container of table salt has a volume of 0.76 L, so the cubic meter could hold approximately 1,316 boxes of table salt!

5. The concentration of trace elements in the Earth's crust; chemical compounds in forensics; dissolved minerals and gases in water; and environmental pollutants in the air, water, or soil are all commonly measured and legally monitored in ppm (sometimes expressed as mg/L) or ppb (sometimes expressed as ug/L). For example, research on greenhouse gases and global warming indicates that since the beginning of the industrial revolution, the concentration of carbon dioxide in the atmosphere has risen from about 275 ppm to 387 ppm. Environmentalists argue that the safe upper limit is no more than 350 ppm. See Activity #20 in this book to discover how something as simple as the lightbulbs we use can make a difference.

Answers to Questions in Extensions, #2 and #3

2. The logic sequence goes as follows: Popular Chinese soups (e.g., won-ton) are primarily water, which has a density of 1 g/ml (= 1 g/cm^3) at room temperature. So, 1 L of water has a mass of 1,000 g, or 1 kg. A cubic meter (or 1 m^3) contains 10 L × 10 L × 10 L "boxes," which equals 1,000 L. One thousand liters (1,000 L) would have a mass of 1,000 kg, or 1 metric ton. *Note:* Because a 1 kg mass "weighs" about 2.2 lb., 1 metric ton "weighs" approximately 2,200 lb.!

3. Laundry liquid detergents contain 70–80% water. Given the relatively high mass per unit volume of water (1 g/ml), it takes a significant amount of energy to move water from the manufacturer to the users (who obviously have ample supplies of water at the local site of use if they are using washing machines). Also, for the equivalent amount of cleansing power (and number of loads of laundry/dollar amount), liquid-based products will require much more packaging material (which, in the case of the liquid detergents, will be plastics made from nonrenewable petroleum instead of renewable paper products for the powdered detergents). This is one of many win-win examples where going green and being good for the environment make economic sense for the consumer.

Section 4: Science-Technology-Society (STS) and Real-World Science Instruction

Medical Metaphor Mixer
Modeling Infectious Diseases

Expected Outcome

Learners introduce themselves to "future friends" and exchange some clear, colorless liquid from their respective plastic cups. They subsequently discover that some of them "test positive" when the addition of an acid-base indicator causes their liquid to turn pink.

Science Concepts

This activity is a *visual analogy* or *simulation model* for the spread of infectious diseases by the exchange of bodily fluids (e.g., sexual contact and airborne sneezes). Infectious diseases kill more people worldwide than any other single cause. The microworld of the very small (i.e., bacteria and viruses) can have a large impact on our everyday macroworld. "Out of sight" cannot be "out of mind" when it comes to microbes and our health. (*Note:* Although the chemistry of this reaction is not the focus of this activity, interested teachers can refer back to Activity #12 for more information about indicators and acid-base reactions.)

Science Education Concepts

Some grade 5–12 lab activities that culture bacteria found around the school environment pose unnecessary safety risks. Inquiry activities that use safe, noninfectious microbes (e.g., milk-souring, yogurt-forming, and compost-degenerating bacteria or sugar-fermenting yeast) are better alternatives (see also Extension #4, p. 230 for virtual diseases). In addition, *analogical model systems* provide safe simulations for teaching topics that are essential for both personal, individual well-being and broader science-technology-society (STS) community health implications.

This activity also serves as a fun Engage-phase mixer that gets learners out of their seats and directly relates to the topic for the unit. hands-on exploration (HOE) is not restricted to formal, period-long activities that must take place in a laboratory space. This activity serves as an Engage-phase mixer that gets learners out of their seats and thinking about a potentially sensitive, socially relevant topic in a fun, low-key way. Ignorance of the science of HIV transmission and AIDS prevention can result in far more serious consequences than losing points on a paper-and-pencil test. More broadly considered, this activity can be used with teachers as a *visual participatory analogy* for highlighting the following science education ideas: Some types of hands-on learning can happen in regular, nonlaboratory classroom settings in shorter time intervals than full or double period classes; hands-on exploration (HOE), discrepant-event experiences challenge students to "turn over the fields of their minds" (or prior conceptions) in preparation for planting "new seeds" (or, more scientifically, accurate conceptions); and there is a "contagious" nature to the teachers-helping-teachers networks in spreading best practices.

Medical Metaphor Mixer

Materials

- Each learner will need a numbered, clear, colorless plastic cup that is filled approximately one-quarter with water. One or two of the numbered cups should be "doped" with one-half of an eyedropper of concentrated (3–6 M) sodium hydroxide solution.
- Solutions of other solid bases (e.g., calcium hydroxide) will also work. The concentration is not especially critical, but do not use ammonia due to its recognizable odor.
- One or more small dropper bottles of phenolphthalein are needed as a base indicator.
- *Optional:* Background music, including songs from different eras—such as "Getting to Know You" (*The King & I* musical); "I Want to Hold Your Hand," "Hello, Goodbye," or "In My Life" (Beatles); and "Who Are You" (The Who)—can be used for effect and to help set a time limit on the activity.

Safety Note

Learners should be informed *before* the cups are distributed that unknown liquids should be treated as potentially dangerous; they should not be touched, tasted, or spilled.

Points to Ponder

Microbiology is usually regarded as having no relevance to the feelings and aspirations of the man of flesh and blood ... It is not only because microbes are ubiquitous in our environment, and therefore must be studied for the sake of human welfare. More interesting, and far more important in the long run, is the fact that microbes exhibit profound resemblances to man.

—Rene J. DuBos, French American bacteriologist (1901–1982)

It is not usually recognized that for every injurious or parasitic microbe there are dozens of beneficial ones. Without the latter, there would be no bread to eat or wine to drink, no fertile soils and no potable waters, no clothing and no sanitation. One can visualize no forms of higher life without the existence of microbes. They are the universal scavengers. They keep in constant circulation the chemical elements which are so essential to the continuation of plant and animal life.

—Selman Abraham Waksman, American microbiologist (1888–1973)

Procedure

1. Have each learner pick up a numbered cup of an unknown liquid after you make this safety announcement: Do not taste, touch, or smell! If desired, turn on the *optional* background music for atmosphere.

2. Instruct the learners to walk up to a peer and pour about half of their liquid into his or her cup, then have the peer reciprocate so that both cups contain both liquids and approximately the same final volume as they initially contained.

When Working With Teachers

If this activity is used at the start of a professional development program, have the teachers introduce themselves to one another (e.g., one of my strengths as a science teacher is ___ , a hobby that recharges my batteries is ___ , a topic or unit that I struggle the most with is ___) or exchange answers to one or two questions related to the purpose of the program (e.g., What do you personally hope to get out of this session? What are the main costs or barriers versus the benefits to incorporating this innovation in my classroom?).

3. Members of the initial dyad should separate and then exchange fluids with two different partners. If the group size is large (more than 20 participants), this pairing and sharing can continue for an additional cycle if desired.

4. After everyone has returned to their seats, ask:
   a. What do you think we are modeling with this activity?
   b. Could you tell at the start of the activity that one or two of your peers were "infected" with a disease-causing microbe?
   c. What would we need to determine who is infected now?

   Pass around one or more dropper bottles of phenolphthalein for everyone to place 1 or 2 drops into their own individual cups. Anyone who exchanged fluids with either an initially infected individual or a secondarily infected individual will have his or her liquid turn pink (due to the presence of the base or "infectious agent").

5. If time permits, use the cup numbers and individuals' recollections of who they partnered up with to find the originally infected person(s), modeling the work of epidemiologists.

Debriefing

When Working With Teachers

Discuss the importance of safe but memorable activities that let science do the talking rather than the teacher just telling the facts or preaching about the dangers as related to STS, health-related issues that involve personal behaviors that can put students at risk. Many students often perceive science, especially constructs that cannot be directly seen or seem to be so different in scale and form from humans (i.e., microbes), as being irrelevant to their lives (see the DuBos quote). Practicing good health and hygiene (e.g., washing your hands with soap before eating and after using the restroom; not placing your hands in your eyes, mouth, or nose; not sharing drinks; directing a sneeze into a tissue or at least down and away from people; abstaining from sex or practicing safe sex) is clearly most relevant. This activity also provides an opportunity to emphasize the unity that underlies the immense diversity of all forms of life. There is a direct line of evolutionary "descent with modification" from the earliest forms of life, the prokaryotes that originated approximately 3.8 billion years ago, to the "newcomers," humans who emerged around 100,000–200,00 years ago. If used as a mixer activity when working with teachers who do not already know each other, take time to share some of their getting-to-know-you questions and answers and to emphasize the power of teacher networks to share and spread best practices.

When Working With Students

This activity serves as an Engage-phase introduction to a unit on infectious disease, including sexually transmitted diseases such as HIV and AIDS. This unit could also be used more broadly to emphasize the positive roles played by microbes that, as Waksman's quote indicates, typically get a bad rap (for examples of positive roles, see Wikipedia under Internet Connections or search Google for topics such as bioremediation, decomposition, fermentation, and probiotics); feature the

powers-of-ten scale differences in size between eukaryotic cells, prokaryotic bacteria, and viruses (e.g., the human body actually contains more bacterial cells than human cells—see Activity #17, Extension #1, such as the Altoona List of Medical Analogies, Biology); and discuss the use of analogies in science and medicine (see Internet Connections on pp. 232–234). Many STS topics lend themselves to problem- (or project-) based learning (see IMSA website in the Internet Connections) where students use the school and larger community as their laboratory (see examples of such activities in Extension #5). PBL team projects such as community surveys; awareness building for community health and/or eco-wise, environmentally friendly programs; and service-learning initiatives allow students to learn science, use skills of active civic engagement, and potentially have a positive social impact rather than merely prepare for a test and earn a course grade.

Extensions

1. *Black-Light Magic and Medical Mixers*: UV-active Glo Germ powders (or lotions) can be exchanged by shaking hands with "infected" individuals. Such powders are sold commercially as "ultraviolet theft detection powders" that are invisible under normal lighting but fluoresce under UV (e.g., see *www.electromax. com/thiefpowders.html* and the Educational Innovations, Kansas City Public Television, and Steve Spangler Science websites in Internet Connections). An alternative, low-cost simulation can be run by rubbing peppermint extract on the palm of one student, who initiates a chain reaction of handshakes. Given the sensitivity of the human nose to even very low concentrations of some chemicals, the fifth or sixth student in a chain will still be able to pick up a detectable smell from his own hand (see also Activity #17, Extension #5 on serial dilution).

    Relate this finding to the public health codes about employee hand washing that are posted in restrooms in restaurants. Germs may be out of sight, but they should not be out of mind: What you can't see *can* hurt you. Keep your hands and germs to yourself. This hands-on activity can be used as an entry point to many topics, such as the variety of germ-caused diseases that can be transmitted by touch and the relative effectiveness of soap and water, hand sanitizers, and antibacterial soaps (see Mayo

Medical Metaphor Mixer

Clinic website); the rise of antibiotic-resistant bacteria due to misuse and overuse of antibiotics; the historical development of the microscope and its role in opening up new worlds of discovery without the need to travel to foreign lands; and the roles of individuals and organizations (e.g., Centers for Disease Control and Prevention and World Health Organization; see the Internet Connections) in containing the spread of infectious diseases and promoting human health as a basic human right.

2. *Science Fiction Film Fun*: Several popular science fiction books and movies have featured outbreaks of highly infectious diseases (e.g., *The Andromeda Strain, Outbreak, Mission: Impossible III*) caused by "bad" micro-organisms. It is harder to find films that feature microbes as "the good guys." One example is the H.G. Wells classic novel *The War of the Worlds* (1898), which was turned into a radio broadcast (by Orson Welles in 1938), and movie, where the Earth's bacteria save humanity by killing off the Martian invaders, whose immune systems aren't "tuned" to fight the bacteria that evolved on Earth. Short segments of movies such as these can be used to initiate discussions about science facts behind the fiction (see, for example, the Blick on Flicks feature in *NSTA Reports* in the Internet Connections). Similarly, a search of online booksellers will uncover a variety of books that discuss the science "facts" in relationship to science fiction stories and films.

3. *Microbes Matter in Human History*: Although history classes often focus on wars, the molecular "arms race" and endless battles between infectious agents (i.e., viruses, bacteria, and fungi) and the immune systems of their human hosts are constant scientific subtexts to all of history. Our immune systems "learn" both within individual humans (e.g., production of antibodies) and evolutionarily as a species across time. But the rapid reproduction rate and short generational time of our microbial enemies mean our victories (e.g., the development of antibiotics) are often short-lived. History textbooks typically give little if any attention to the role of microbe-induced diseases in shaping human civilizations. Consider developing a collaborative unit with history teachers to discuss some of the ideas found in classic works in this field:

- Buckman, R. 2003. *Human wildlife: The life that lives on us*. Baltimore, MD: John Hopkins University Press. "Your body

has 100 trillion cells, but only 10 trillion are human. The rest belong to the bacteria, fungi, viruses, and parasites that live on or in us" (from book's back cover).
- Cartwright, F. F., and M. D. Biddiss. 2000. *Disease and history.* 2nd ed. New York: Dorset Press. This book covers the period from ancient Greek civilization to the fall of the Russian monocracy in 1917.
- Crawford, D. H. 2009. *Deadly companions: How microbes shaped our history.* New York: Oxford University Press.
- Diamond, J. 1997. *Guns, germs, and steel: The fate of human societies.* New York: W.W. Norton. Thirteen thousand years of world history are viewed through the lenses of anthropology, behavioral ecology, epidemiology, and technological development. For instance, Diamond argues that many more New World natives were killed by the diseases carried by the European invaders than by their superior weapons. See the National Geographic Society's film in the Internet Connections.
- Dixon, B. 1994. *Power unseen: How microbes rule the world.* New York: W.H. Freeman. This book features 75 microbes that have affected, and continue to affect, history.
- McNeill, W. H. 1977. *Plagues and people.* New York: Anchor. McNeill offers an interpretation of world history as seen through the political, demographic, ecological, and psychological impacts of disease on cultures, from the conquest of Mexico by smallpox to AIDS in the early 1980s.
- Neiss, R. M., and G. C. Williams. 1994. *Why we get sick: The new science of Darwinian medicine.* New York: Vintage Books/Random House.
- See also the National Center for Case Study Teaching in Science website in Internet Connections.

4. *Virtues of Virtual Diseases*: Another safe, interactive, and inquiry-based way to study the sources and spreading mechanisms of disease is by way of multi-user virtual environments such as the River City site (for middle school) and online gaming environments such as the annual virtual epidemics of WhyPox and WhyFlu on Whyville (for preteens). See Internet Connections.

5. *Science for Life and Living*: Whether or not teachers are using a textbook that features science-technology-society (STS) issues,

Medical Metaphor Mixer

finding relevant STS issues to infuse into the science curriculum is as easy as scanning the headlines of the local and national news. The student question "So what, who cares?" (SW_2C) is not impertinent but most pertinent, and it deserves a better answer than "You need to know the science content for the next test, college, or a future career."

Challenge teachers or students to brainstorm a list of science issues in the news that directly relate to the science content and skills currently being taught. The lists in the chart below include some possibilities. Analyzing (much less offering viable solutions for) real-world STS issues demands multi- and interdisciplinary connections across the sciences, as well as knowledge and skills that students acquire in English/language arts, mathematics, social studies, and technology classes. This fact opens up the possibility of units and projects that cut across curriculum areas and transform the school into a learning organization and the surrounding community into a laboratory for learning, serving, and making a difference. As science becomes more than just another academic subject, "think globally and act locally" becomes more than a slogan.

Biology ←→	Chemistry ←→
AIDS and other STDs	Agricultural chemicals and food production
Antibiotics (overuse and misuse) and probiotics	Alternative fuels, batteries, and energy sources
Biodiversity and ecological health	Cosmetics and personal cleanliness
Cancer testing, treatment, and prevention	Food additives and supplements
Diet, nutrition, health, and longevity	Fossil fuels and petrochemicals
Epidemiology and bioterrorism	Household chemicals, safety, and environment
Evolution versus religion	Manufacture and cradle-to-grave cycle of synthetics
Genetic testing and engineering	Materials science and engineering
Pharmaceuticals, drugs, and vaccines	Pollution abatement and remediation
Population and reproductive technologies	Recycling and waste disposal
Earth Sciences ←→	**Physics** ←→
Climate modification	Communication and computer technologies
Earthquake prediction and safety	Electromagnetic radiation and health concerns
Flooding, forest fires, and disasters	Energy conservation and efficiency
Fossil fuel: location and mining	Fuel cell and solar energy research and development
GPS, remote sensing, and privacy	Information processing, computing, and security
Groundwater use and contamination	Medical imaging and radiation treatment
Meteorology: prediction and safety	Nanotechnologies
Natural resource mining	National defense and nuclear proliferation
Radioactive waste disposal	Pollution: heat, light, nuclear radiation, and sound
Space exploration and exploitation	Sports science and training technology
	Transportation efficiencies and options

Internet Connections

- Access Excellence: search for disease: *www.accessexcellence.org/AE*
- All the Virology on the www: *www.virology.net* (e.g., *Big Picture Book of Viruses*)
- Altoona List of Medical Analogies: *www.ltoonafp.org/analogies.htm*
- American Society for Microbiology: *www.asm.org*. Search for Education Resources: K–12 Teachers: Classroom Activities: *Outbreak! Investigating Epidemics* (paper exchange simulation) and *The Role of Microorganisms in the Ecosystem* (decomposition experiments).
- Bacteria: More than Pathogens: *www.actionbioscience.org/biodiversity/wassenaar.html*
- Blick on Flicks feature of *NSTA Reports*: *www.nsta.org/blickonflicks*
- Cells Alive: How big is a … ?: *www.cellsalive.com/howbig.htm*
- Centers for Disease Control and Prevention: Infectious Disease Information Index: *www.cdc.gov/ncidod/diseases*
- Cornell Institute for Biology Teachers: Classroom Resources: Everyday Biology: HIV Testing and HIV Cocktail downloadable lab simulations: *http://cibt.bio.cornell.edu/labs/eb.las*
- Discovery Health: How AIDS Works: *http://health.howstuffworks.com/diseases-conditions/infectious/aids.htm*
- Disney Educational Productions: *Bill Nye the Science Guy: Germs* ($29.99/26 min. DVD): *http://dep.disney.go.com*
- Educational Innovations: Glo Germ Classroom Kit and related products (888-912-7474): *www.teachersource.com/BiologyLifeScience/Germs/GloGerm.aspx*
- Fankhauser's Cheese Page (and yogurt): *http://biology.clc.uc.edu/Fankhauser/Cheese/Cheese.html*
- Kansas City Public Television's e-Eats program: *Keep Your Hands to Yourself* (PDF download): *www.kcpt.org/eats* (*Note:* This activity uses Glo-Germ material cited in Education Innovations, above.)
- Illinois Mathematics and Science Academy Center for Problem-Based Learning: PBL resources: *http://pbln.imsa.edu/index.html*

Medical Metaphor Mixer

- *Intimate Strangers: Unseen Life on Earth*: www.pbs.org/opb/intimatestrangers
- Mayo Clinic: Germs: www.mayoclinic.com/health/germs/ID00002

    Hand-washing: Do's and don't's: www.mayoclinic.com/health/hand-washing/HQ00407

    Infectious diseases: www.mayoclinic.com/health/infectious-diseases/DS01145

- Microbe World: www.microbeworld.org
- Microbiology K–12 Curriculum Outline: www.science-projects.com/classexpts.htm#connections
- National Association of Biology Teachers Position Statements: www.nabt.org/websites/institution/index.php?p=35

    The Role of Biology Education in Addressing HIV and AIDS

    The Use of Human Body Fluids and Tissue Products in Biology Teaching

- National Center for Case Study Teaching in Science: http://sciencecases.lib.buffalo.edu/cs

    Abracadabra: Magic Johnson and Anti-HIV Treatments

    The Case of a Tropical Disease and Its Treatment (Chagas): Science, Society & Economics

    A Case Study Involving Influenza and the Influenza Vaccine

    The Mystery of the Blue Death (cholera): A Case Study in Epidemiology and the History of Science

- National Geographic Society: *Guns, Germs and Steel* (2-disc DVD; #93008/$39.95): http://shop.nationalgeographic.com
- National Science Teachers Association Position Statement: The Teaching of Sexuality and Human Reproduction: www.nsta.org/about/positions/sexuality.aspx
- PBS *Evolution*, Show #4 The Evolutionary Arms Race: www.pbs.org/wgbh/evolution/about/show04.html
- River City (investigate virtual disease): http://muve.gse.harvard.edu/rivercityproject

- Science Buddies: Making yogurt from whole milk: *www.sciencebuddies.org/mentoring/project_ideas/MicroBio_p010.shtml*
- Serendip: Hands-on Activities for Teaching Biology to High School or Middle School Students: *http://serendip.brynmawr.edu/sci_edu/waldron*. Search for Human Physiology (Sexual Health and Reproduction) and Evolution and Diversity (Some Similarities Between the Spread of an Infectious Disease and Population Growth)
- Steve Spangler Science: Cover Your Mouth; Germ Science (lotion glows under UV light): *www.stevespanglerscience.com/experiment/glo-germ-and-giantmicrobes*
- U.S. National Library of Medicine and the National Institutes of Health, Medline Plus: Infectious Diseases: *www.nlm.nih.gov/medlineplus/infectiousdiseases.html*
- Whyville (site of WhyPox and WhyFlu virtual epidemics): *www.whyville.net/smmk/nice*
- WikiPedia: *http://en.wikipedia.org/wiki*. Search topics: AIDS, bioremediation, decomposition, fermentation (food), germ theory, HIV, infectious disease, influenza pandemic, Methicillin-resistant Staphylococcus aureus, and probiotics
- World Health Organization: HIV/AIDS: *www.who.int/hiv/en*

Answers to Questions in Procedure, step #4

4. a. Someone will probably suggest that the activity is a model for the spread of STDs in general or HIV and AIDS in particular. Clarify that other infectious diseases can also be spread through direct physical contact, breathing air that still contains viruses suspended on droplets from a sneeze, and so on. Many of the major events in human history are associated with the spread of microbe-induced infectious diseases (see Extension #3).

Medical Metaphor Mixer

b. No, at the start of the activity, no one showed any obvious symptoms of being infected or ill, and all the cups of fluid looked the same.
c. Some kind of chemical test or biological assay would be needed, in this case an acid-base indicator.

Activity 19

Cookie Mining
A Food-for-Thought Simulation

Expected Outcome

Various brands, sizes, and textures of chocolate chip cookies are "mined" as an edible analogy for natural resource management and conservation.

Science Concepts

Environmental and Earth science concepts such as mining the Earth's limited supplies of unevenly distributed, nonrenewable but recyclable resources; conservation; and ecology/economy tradeoffs can be introduced.

Science Education Concepts

This activity is a quick model of an *analogy-based simulation* type of hands-on exploration (HOE) that uses readily available, relatively inexpensive, safe materials to introduce an important science-technology-society (STS) issue. It can be used as an Engage-phase introductory activity for a unit on resource management and environmental conservation, or as a "food for thought" break in a class or workshop with teachers.

Acknowledging learners' nutritional needs with a surprise snack has both practical and socially symbolic significance (even if the snack is occasionally a higher-calorie, lower-nutrient cookie). If learners are hungry, they are not able to focus on learning, and when people break bread together, they can create community bonds. And who can resist a science class that treats learners to a tasty treat that feeds both the mind and the body! It is also worth noting that humans require regular inputs of carbohydrates, fats, and protein sources of molecular energy to fuel our cellular "fires," analogous to human-designed machines' use of hydrocarbons. Mining fuels and minerals can dramatically affect the environment. If desired, inquiry-oriented science teaching can be discussed as "digging into discrepancies" and "mental mining" for "nuggets of gold" (i.e., scientifically valid conceptual understanding).

Materials

- 3–5 varieties of chocolate chip cookies that vary in size, texture, and amount and type of chocolate chips, and other "resources" (e.g., walnuts)
- Paper towels and toothpicks for use as "mining tools"
- Expanded versions of this simulation include economic variables (see Internet Connections).

Safety Note

"Materials intended for human consumption shall not be permitted in any space used for hazardous chemicals and or materials." NSTA Position Statement *Safety and School Science* (*www.nsta.org/about/positions/safety.aspx*). Therefore, this activity should be done in a classroom on desktops or tables, not in a formal lab setting.

Cookie Mining

Points to Ponder

Aluminum was once a precious metal.
—Jules Verne, French writer (1828–1905)

We travel together, passengers on a little space ship, dependent on its vulnerable reserves of air and soil ... preserved from annihilation only by the care, the work, and I will say, the love, we give our fragile craft.
—Adlai Stevenson, American politician and presidential candidate (1900–1965)

UNLESS someone like you cares a whole awful lot, Nothing's going to get better. It's not.
—The Once-ler character in Dr. Seuss's (Theodore Geisel, 1904–1991) classic 1971 book (and animated movie) *The Lorax*

Procedure

Introduce the activity as an analogy for mining and interject the questions listed below and their answers (pp. 245–246) as the activity proceeds. Tell the learners to refrain from eating any part of their "lab materials" and that you will give each of them their own "resources" at the end of the activity.

Safety Note

It is easy to spread germs if multiple learners' hands are on any given cookie, so the cookies that are "mined" should not be consumed.

1. Group the learners in teams of between two and seven learners. Distribute two cookies (one as a control) to each team, using a variety of brands for different teams. If the cookie was used as an analogical model to represent the Earth's crust, what might the chocolate chips, walnuts, and other ingredients represent? What might the different size groups represent?

2. In looking around the room and comparing your chocolate chip cookie to those of other teams
   a. What differences are immediately evident?
   b. How might these differences relate to the Earth's resources (e.g., metals obtained from minerals and fossil fuels)?

3. Are chocolate chips a renewable or nonrenewable resource? Given their standard use (as food), are they recyclable or nonrecycable? How are chocolate chip cookies similar to or different from actual mineral resources with respect to these questions?

4. Give each team several toothpicks and challenge them to compete to remove as many chocolate chips from one of their two cookies as they can in the next 2 minutes.

5. If the chips are the most important part of the cookie and you extracted and consumed all of your chips, what might you do to obtain more chips? How does this relate to human societies and history?

6. Examine the cookie remains relative to your control cookie. How is chocolate chip mining similar to and different from actual mining? How easily can you restore the mined area to its natural state? Why is land reclamation important?

7. How would we need to modify the mining rules and objectives to better preserve the quality of the cookie remains? Would some of the chips need to be left unmined?

8. What tradeoffs are involved in mining? How does recycling metals, glass, and plastics help reduce some of the difficult choices facing society?

9. Distribute fresh cookies for the learners to eat. Are the chips more like metallic resources (e.g., aluminum, iron, copper) or fossil fuels?

10. Share the three quotes and mention that prior to modern electrochemical separation processes, aluminium, one of the most abundant metals in the Earth's crust, was more expensive than gold (it now is used as throwaway foil). Ask: What questions do these quotes and this activity raise for you about our use (and abuse) of the Earth's resources? Even though the amount of any nonrenewable metallic or fossil fuel resource on our planet is a constant, how do new technologies effectively increase the supply of fixed resources?

Debriefing

The basic analogy is discussed as the activity proceeds (see Procedure).

When Working With Teachers

Discuss the value of analogies and simulations, simple hands-on explorations (HOEs), and incorporation of STS content and environmental education into the science curriculum to help develop scientifically literate, environmentally responsible citizens. Teachers can review the relevant NABT and NSTA position statements (see Internet Connections). For teachers who wish to regularly infuse STS issues into their teaching, the following books are sources of up-to-date science content and balanced political perspectives.

- Easton, T. A. 2009. *Taking sides: Clashing views on science, technology, and society.* 9th ed. Dubuque, IA: McGraw-Hill. See also Easton, T. 2008. *Taking sides: Clashing views on environmental issues.* 13th ed. Dubuque, IA: McGraw-Hill.
- Environmental Literacy Council (ELC) and the National Science Teachers Association (NSTA). 2007. *Resources for environmental literacy: Five teaching modules for middle and high school teachers.* Arlington, VA: NSTA Press.
- Miller, G. T. Jr., and S. Spoolman. 2008. *Living in the environment: Principles, connections, and solutions.* 16th ed. Pacific Grove, CA: Brooks/Cole.

When Working With Students

The activity could be used as a quick Engage-phase lesson to introduce a unit on Earth's resources and the importance of environmental conservation legislation and adherence to the 3 *Rs* (*reduce, reuse,* and *recycle*) by individuals and communities. The law of conservation of matter and the boundaries of the Earth suggest that there really is no way to truly throw things away. It is also foolish to waste energy by continuing to throw away metals that require large amounts of energy to mine and refine (and generate pollution in the process). Instead, waste metals can be separated from trash (by hand and other physical or chemical processes) and recycled at a savings to both the environment and the economy.

Extensions

1. *Simulations of Population and Resource Distribution and Use*: Teams can obtain numerical data for a specific country's land area; the percentage of the world's known reserves of coal, oil, and natural gas found in the country; and its population to run more involved, data-based, quantitative versions of this simulation. Consider how the development of automobiles increased the demand for petroleum and shifted the balance of power on the world stage. See the Internet Connections for examples of population growth simulations and animations.

2. *Petroleum Prospecting: A Food-for-Thought Simulation*: Foods such as jelly or cream-filled donuts or cupcakes can be used to model sampling techniques used by prospectors to estimate the depth, breadth, and 3D shape of oil or gas fields (e.g., see Cupcake Core Sampling in Internet Connections). From issues of discovery and mapping, students can move on to explore the environmental and economic tradeoffs of the various means of extracting oil and natural gas from land or offshore (see Internet Connections: Wikipedia). The 2010 BP oil spill in the Gulf of Mexico and the development of the natural gas reserves of the Marcellus Shale in Pennsylvania and New York make for compelling STS case studies.

3. *Bumper Sticker Science*: Have students collect, critically analyze, and create car bumper stickers, public service announcements, or ad slogans that promote environmental literacy and eco-friendly behaviors by using clever, pithy, and scientifically accurate plays on words. Examples of the latter include "Waste is a Wonderful Thing to Mind"; "Mind and mine your waste"; "Recycle—those who ignore nature are condemned to deplete it"; "Mind your mines"; and "Renewable energy is homeland security." Student problem-based learning teams can study various aspects of their own homes and school (e.g., overall production of waste, extent of recycling, environmental friendliness of the chemicals used for cleaning, use of recycled products, energy efficiency of heating and cooling, insulation) and work with their peers, teachers, and administrators to move their homes and school toward being more green, eco-friendly, and economically conscious. Think

globally, act locally! See Internet Connections: Wikipedia: life cycle assessment or cradle-to-grave analysis.

4. *Research and Resource Replacement*: The use of nitrates for explosives and fertilizers is an especially compelling case study of the mining of a differentially distributed natural resource (i.e., nitrates) and the subsequent depletion, pollution, and international trade (and aggression), as well as the impact of new science and technology. This story and the eventual replacement of naturally occurring nitrates with universally available, artificially "fixed" atmospheric nitrogen also played a critical role in both World War I and World War II, as depicted in the book *The Alchemy of Air* (2008) by Thomas Hager. See also Extension #2 in Activity #12.

Internet Connections

- AAAS Atlas of Population and Environment: *http://atlas.aaas.org/index.php?part=2*
- Chris Jordan, Photographic Arts: Running the Numbers: An American Self-Portrait (interactive art displays the statistics of American life; many of the pieces of art relate to environmental issues; see, for example, "Oil Barrels," which depicts 28,000 42 gal. barrels, the amount of oil consumed in the United States every two minutes (equal to the flow of a medium-size river): *www.chrisjordan.com/gallery/rtn/#oil-barrels*; and "Cans Seurat" (2007), which depicts 106,000 aluminum cans, the number used in the United States every 30 seconds: *www.chrisjordan.com/gallery/rtn/#cans-seurat*
- Cookie Mining and Economics: Simulations and Games
    www.earthsciweek.org/forteachers/cookiemining_cont.html
    www.teachcoal.org/lessonplans/cookie_mining.html
- Cupcake Core Sampling: *www.coaleducation.org/lessons/wim/4.htm*
- Environmental Defense Fund: *www.edf.org/home.cfm*
- Environmental Literacy Council: *www.enviroliteracy.org/teachers-index.php*

- Illinois Mathematics and Science Academy Center for Problem-Based Learning: PBL resources: *http://pbln.imsa.edu/index.html*
- National Association of Biology Teachers Position Statements: *www.nabt.org/websites/institution/index.php?p=35*

    Sustainability in Life Science Teaching

    Teaching About Environmental Issues
- National Center for Case Study Teaching in Science: Watch Your Step: Understanding the Impact of Your Personal Consumption on the Environment: *http://ublib.buffalo.edu/libraries/projects/cases/footprint/footprint.html*
- National Science Teachers Association Position Statements on:

    Environmental Education: *www.nsta.org/about/positions/environmental.aspx*

    Teaching Science and Technology in the Context of Societal and Personal Issues: *www.nsta.org/about/positions.societalpersonalissues.aspx*
- Population Action International: Search: Mapping the Future of World Population: *www.populationaction.org*
- Population Connection: World Population Video/DVD and other curriculum resources: *www.populationeducation.org*
- Population Statistics and Resources: *http://geography.about.com/od/obtainpopulationdata*
- Union of Concerned Scientists: Citizens and Scientists for Environmental Solutions: *www.ucsusa.org*
- USGS Mineral Resources Program: *http://minerals.usgs.gov*
- WikiPedia: *http://en.wikipedia.org/wiki*. Search topics: hydraulic fracking, life cycle assessment, mining, and offshore drilling.
- World Population: A Guide to the WWW (links): *http://tigger.uic.edu/~rjensen/populate.html*
- World Resources Institute: Working at the intersection of environment and human needs: Many publications, including *http://water.wri.org/worldresources2005-pub-4073.html*
- World Resources Simulation Center: *http://wrsc.org*

Cookie Mining

Answers to Questions in Procedure, steps #1–#10

1. The chips could represent fossil fuels, mineral ores, or other valuable resources. The different-size groups represent the different population numbers (and/or densities) in different countries.

2. a. The cookies differ in size, color, and texture of the dough and the type, size, number, and location of items embedded in dough.

   b. The variations between the different cookies could serve as a model for the unequal distribution of different types of resources, different soil types, and so on, as well as different size land masses relative to the population sizes.

3. Chocolate chips are a product made from renewable plants, but they are typically consumed and are not recyclable. Metals are nonrenewable and recyclable, and fossil fuels are nonrenewable and nonrecyclable.

4. No question is asked.

5. Teams could trade other resources they have for resources they lack; alternatively, they could raid the supplies from another team. Human history has been driven in part by uneven natural resource availability and the balance between civil, mutually beneficial economic exchanges of resources and wars of aggression to seize resources without compensation. What is seen as a valuable resource typically changes with new science and technologies that enable us to obtain, process, and use the Earth's unevenly distributed resources.

6. Buried or otherwise hidden resources must be found and "mined," which results in some disruption of the land and pollution of the environment and can require large quantities of energy. A cookie is not a complex living system like the Earth's biosphere, and it has no means of restoring itself once it is mined. The ease of restoring mined land depends on how the mining operation was conducted. Strip mining that was once allowed in the Appalachian mountains proved to be an environmental disaster. Land reclamation is essential to controlling erosion, preserving

the quality of groundwater, and maintaining biodiversity and natural beauty. These efforts should be seen as both a cost of doing business and an investment for the future.

7. The rules of the game (or environmental laws) could be changed from obtaining the most chips in the least amount of time to including a requirement for land preservation and reclamation. This might involve leaving some chips underground until better technologies become available.

8. Tradeoffs include the energy costs and environmental pollution produced versus obtaining minerals necessary for commerce at a competitive economic rate. Recycling previously mined minerals that have already been processed into products typically results in both a savings of energy (i.e., it takes more energy, and therefore more money, to mine the raw ore) and a reduction of pollution—this is a win-win for the environment and the economy. Also, if the raw material is imported from another country, our trade balance and national security may be affected positively or negatively.

9. Unlike either metals or fossil fuels, chips are renewable since they are continuously being made by plants. They are nonrecyclable like fossil fuels because they are "burned" (oxidized in cellular respiration) for energy. Metal can and should be recycled whenever possible, rather than allowed to oxidize in landfills and pollute our water resources.

10. The first question is one for open-ended discussion. With respect to the second, athough new technologies constantly extend the limits of exploration and minable depths and extraction capabilities from lower concentrations, they typically do so at some increased economic and environmental costs (e.g., deep-sea oil drilling and the BP Gulf of Mexico oil spill in 2010). Ultimately, the 3 Rs of conservation (reduce, reuse, recycle) tend to make sense from both environmental and economic perspectives (see Activity #20).

Making Sense by Spending Dollars

An Enlightening STS Exploration of CFLs, or How Many Lightbulbs Does It Take to Change the World?

Expected Outcome

The wording and validity of the scientific and economic claims for compact fluorescent lightbulbs (CFLs) are critically examined. Their environmental impact relative to standard incandescent lightbulbs is also discussed.

Science Concepts

This activity can be used to introduce a range of topics, including the use and misuse of scientific units of energy (kilowatt-hours), power (watts), and light output (lumens); the stoichiometry of fossil fuel combustion reactions; and the mathematics of environmental economics and consumer purchases. The *discrepancies* that are investigated include the following: A lightbulb that is more expensive to purchase actually saves consumers a significant amount of money over time (due to its greater energy efficiency and longevity), which makes both environmental and economic sense; and scientific terms are sometimes inappropriately used on product packaging, which creates misconceptions that confuse distinct concepts.

Science Education Concepts

The validity of "scientific" claims in advertising is an important science-technology-society (STS) issue, as is educating citizens to look at the tradeoffs between overall costs, risks, and burdens (e.g., environmental impacts) and benefits (e.g., short- and long-term economics, quality of light) when making consumer choices. For the past 100 years, the incandescent lightbulb has served as the metaphorical symbol for bright, innovative ideas. The contrast of new and improved CFLs to incandescent lightbulbs ("yesterday's best answer") serves as both a *visual participatory analogy* and a *model activity* for real-world curriculum-instruction-assessment (CIA) standards that focus on science for all students (rather than only the elite few). Providing high-quality science-technology-engineering mathematics (STEM) education (NRC 2010) and promoting responsible consumerism and eco-friendly citizenship are synergistic (not mutually exclusive) goals. Scientific literacy matters in everyday life, and individual actions multiplied by millions of people can have a huge environmental impact, either positive or negative. "Science for all Americans" arguments include the need for an involved citizenry that can make informed individual choices and contribute to public debate and decisions to help guide STS-related policies and government regulations, as well as research investments.

Making Sense by Spending Dollars

Materials

- 1 CFGL and 1 standard incandescent lightbulb of either approximately the same light output or same wattage
- Photocopies of one or more manufacturer product claims for CFLs (e.g., Honeywell, Phillips, Sylvania)
- Extension #1 requires multiple equivalent light-output bulbs and electric power meters for an optional measurement-based investigation.

> ### Points to Ponder
>
> *No category of sciences exists to which one could give the name of applied sciences. There are science and the applications of science, linked together as fruit is to the tree that has borne it.*
>
> —Louis Pasteur, French chemist and microbiologist (1822–1895)
>
> *[Edison] definitely ended the distinction between the theoretical man of science and the practical man of science, so that today we might think of scientific discoveries in connection with their possible present or future application to the needs of man. He took the old rule-of-thumb methods out of industry and substituted exact scientific knowledge, while on the other hand, he directed scientific research into useful channels.*
>
> —Henry Ford, American industrialist and auto manufacturer (1863–1947)
>
> **Scientific literacy** *is the knowledge and understanding of scientific concepts and processes required for personal decision making, participation in civic and cultural affairs, and economic productivity … Scientific literacy implies that a person can identify scientific issues underlying national and local decisions and express positions that are scientifically and technologically informed. A literate citizen should be able to evaluate the quality of scientific information on the basis of its source and the methods used to generate it.*
>
> —*National Science Education Standards* (NRC 1996, p. 22)

If desired, Activity #15, "Brain-Powered Lightbulb," in *Brain-Powered Science* can be used to introduce the topic in a humorous, puzzling way. The battery-powered, "magic" lightbulb used in that activity is is available from the following retailers:

- Abracadabra Magic (Miracle Light Bulb): *www.abra4magic.com* (A079/$9.99)
- GagWorks (magic lightbulb): *www.gagworks.com* (#GW1158/$8.99)
- Magic Trick Store: *www.magictrickstore.net/magiclightbulb* ($9.95)

Procedure

When Working With Teachers Only

What do the three quotes in Points to Ponder suggest about the relationship between science and technology and the need for and value of infusing STS and STEM topics into science curriculum-instruction-assessment? For a specific example of the problem with the absence of scientific literacy, consider sharing the following quote from Supreme Court Justice Antonin Scalia (and the audience's reaction) at the November 29, 2006, public hearings on a suit filed by 12 states and the District of Columbia to challenge the EPA's right to regulate emission of greenhouse gases on new cars: "I told you before I'm not a scientist. ... That's why I don't want to have to deal with global warming, to tell you the truth" (*http://en.wikipedia.org/wiki/Massachusetts_v._Environmental_Protection_Agency*. See External Links: Transcripts of Oral Arguments [PDF]). His comment was greeted with smiles and laughter.

When Working With Students (and Teachers)

1. Initiate a discussion* about energy conservation by holding up a CFL and a standard incandescent lightbulb and asking these questions:

    a. Can buying a more expensive compact fluorescent lightbulb (CFL) make sense (or save cents) relative to buying a standard, cheaper incandescent lightbulb (e.g., $3 versus $1)? How?

    b. What information would you need to verify the manufacturer's claim that you could save money by buying their product? What factors affect the total operating cost of a lightbulb?

Making Sense by Spending Dollars

c. Can we conserve electricity (and save money) used in lighting without "being in the dark" more often? Can conservation be convenient, comfortable, and sensible?

Optional introduction: Use the discrepant demonstration of a "magic, brain-powered" lightbulb (compared to a standard incandescent one given to a student volunteer) to establish the need for "bright" ideas to conserve energy, money, and natural resources before revealing the newer CFL, with its creative twist design.

2. For a quick, visual, qualitative comparison, screw a CFL and an incandescent lightbulb with equivalent light-output ratings into identical lamps and turn them on (or, for contrast, screw in two bulbs that have the same wattage and note the different light outputs). Extension #1 offers suggestions for ways to study the two types of bulbs empirically.

3. In groups of three or four, examine the package (or a photocopy of the package) for a compact fluorescent bulb. Make a list of the product claims, paying special attention to numerical data and units of measurement. Is there any information provided on the package that is scientifically incorrect? *Note:* The numbers listed below in step #4 are used as examples; in this step, be sure to use the exact numbers from whatever products you actually use (e.g., an 18 W CFL versus a 75 W incandescent bulb; some CFLs claim a life of 10,000–25,000 hours) and the cost of electricity per kilowatt-hour where you live, which may be as much as 100% higher than the figure used in step #4.

4. Calculate the total operating cost of using the 15 W compact fluorescent versus the standard 60 W incandescent bulb for a period of 6,000 hrs. (light output of each = about 950 lm). Assume that 1 kwh (1 kilowatt-hour is a unit of electricity equivalent to ten 100 W bulbs burning for one hour) of electricity costs $0.10 and that a standard incandescent bulb costs $1. If desired, have students calculate the usage time needed to break even on costs and begin saving money (typically a few months, at most).

5. Besides personal economics, what other personal, societal, and environmental benefits are achieved by using a compact fluorescent versus a standard incandescent lightbulb? What data might be helpful to establish the validity of these benefits?

6. If 8 lbs. of coal (or 0.9 gal. of fuel oil) are used to generate 10 kwh of electricity and 29 lbs. of carbon dioxide are produced in the process, consider asking chemistry students to use stoichiometric ratios to calculate theoretical weight of CO_2 produced (assuming complete combustion of carbon → 1 mol CO_2/1 mol C = 44 g CO_2/12 g C). How many pounds of coal are saved (i.e., not burned) and pounds of CO_2 emissions reduced (i.e., not produced) over the lifetime of one CFL?

7. Are there any potential hidden costs that might be higher for the compact fluorescent than for standard incandescent bulbs? Cost-benefit analyses typically involve the concept of tradeoffs. Viewing all the relevant factors, do you feel CFLs make more sense than incandescent bulbs?

8. What scientific experiments might you conduct to further study the costs and benefits of this new technology?

9. Are long-term environmental and short-term economic concerns always at odds with one another? Cite other examples of win-win, Earth-friendly technologies.

Debriefing

When Working With Teachers

STS and STEM issues can be integrated into nearly every unit of study across the various science disciplines (see Activity #18, Extension #5). Infusing real-world issues into the curriculum motivates interest in STEM, relates otherwise abstract concepts to everyday concerns, and creates educated, environmentally responsible citizens. Such informed, action-oriented citizens can make a difference. Invite teachers to review and discuss the NABT and NSTA positions statements on STS and environmental education (see Internet Connections). Global warming and the role of fossil fuels in producing carbon dioxide as a major greenhouse gas has been acknowledged by scientists, national governments, and concerned citizens (e.g., see carbon footprint calculators in the Internet Connections). This activity shows that we don't necessarily have to sacrifice quality to be environmentally conscious consumers (and, in fact, saving cents and the environment at the same time makes sense).

Making Sense by Spending Dollars

Relevant background information: Historically, the incandescent lightbulb dates back to the 1906 substitution of longer-lasting tungsten filaments into Edison's 1879 development of a 40-hour carbon filament bulb and the 1913 introduction of inert gas to extend filament life by reducing oxidation. Few changes occurred in this basic heat bulb design until the development of more efficient, cool-light fluorescent tubes in the late 1930s. Approximately 90% of the energy input of an incandescent lightbulb is given off as heat, versus 10% as light. In the mid-1980s, Phillips introduced the screw-in "Earth Light" bulb, or CFL, at an upfront cost of $20 but with a valid claim that the consumer would save more than $57 over the extended life of the bulb. Since then, improvements in manufacturing and competition have lowered the price of a compact fluorescent bulb to about $3, so that such bulbs make even more economic sense. Despite CFLs' clear advantages (i.e., they use 75% less energy and last up to 10 times longer than comparable incandescent bulbs), when it comes to lighting, most American households and businesses are still living in the "dark ages." Lighting consumes about 22% of the electricity produced in the United States, so more efficient lights can make a real difference in terms of energy savings and environmental costs (e.g., slowing and reversing the atmospheric concentration of carbon dioxide that has risen since the beginning of the industrial revolution from about 275 ppm to 387 ppm today (see Activity #17 for models of the idea of parts per million). See also the reference books listed in Activity #19.

When Working With Students

Beyond reviewing the eco-math and discussion questions, be sure to emphasize the importance of both smart technologies and educated consumers who make wise choices that are in their own, society's, and the biosphere's long-term best interests.

Extensions

1. *Comparative Consumer Testing*: P3 International manufactures Kill A Watt #P4400, an electronic device that easily measures volts, amps, watts, line frequency (Hz), and kilowatt-hours when the appliance is plugged into the unit, which is then plugged into a

wall outlet, making otherwise invisible energy use visible (*www.p3international.com/products/special/P4400/P4400-CE.html*). Such activities should be performed away from water and with GFCI circuits. "Energy vampires" or the "phantom loads" of devices that use energy even when they are turned off (i.e., in standby, instant-on mode) can account for 5–10% of residential electricity consumption (see Internet Connections: Lawrence Berkeley National Laboratory). Students can quantitatively compare the stated and measured statistics for incandescent bulbs, CFLs, and light-emitting diodes (LEDs). Various online suppliers sell this device for $17–40 (e.g., *http://scientificsonline*, item #X30542-22 and #X30859-96). The Watts Up Power Meter is similar but more expensive. Photographers' light meters can also be used to measure the actual light outputs, as compared to naked-eye qualitative assessments and product claims. Simple, diffraction grating-type spectroscopes can be used to contrast the continuous color spectrum of incandescent bulbs to the bright line spectra of fluorescent tubes, CFLs, and LEDs that are related to electron transitions between different energy levels in atoms. *Note:* LEDs use about half the energy (for a given light output) and last up to 10 times longer than CFLs and do not contain mercury. Research is under way to produce white LEDs that can economically match the light output of CFLs. LEDs are most likely the next-generation bright idea and are already being used in long-lasting flashlights that run on either built-in, hand-powered generators or batteries.

2. *Carbon Footprint Calculators and Global Warming Information on the Internet*: Students can compare the results produced by carbon calculators from various commercial companies (.com), environmental organizations (.org), and government sites (.gov). In a related vein, the larger STS issue of global warming and the greenhouse effect can be discussed in the context of the perspective of different websites and emotion-generating images (e.g., video clips from former vice president Al Gore's film *An Inconvenient Truth* or TV ads from the Environmental Defense Fund; see Internet Connections). Students can explore the scientific uncertainties involved in long-range climate forecasting and the relationships between energy conservation, pollution reduction, international economics, and geopolitical positions.

Making Sense by Spending Dollars

3. *Taking It to the Streets: Marketing "Bright" Ideas in the Home, School, and Society*: STS topics lend themselves to interdisciplinary instruction and curriculum-embedded authentic assessments where students demonstrate their understanding through real-world projects that make a difference (see Internet Connections: Illinois Mathematics and Science Academy Center for Problem-Based Learning). Students can collect data from their parents' electric bills on the average number of kilowatt-hours used in their homes in a typical monthly period and account for differences in terms of variables such as the season, the size of households, whether the house relies on electric heat and/or air conditioning, the types of lightbulbs used, and so on. In the case of CFLs, students could develop an awareness-raising campaign (and/or eco-friendly fund-raiser) to convince consumers to make more use of energy-efficient lighting options in their homes, schools, and local government buildings.

Internet Connections

- AAAS Project 2061: *Communicating and Learning About Global Climate Change: An Abbreviated Guide for Teaching Climate Change* (read this 2007 publication online or order): *www.project2061.org/publications/order.htm#TeachingGuides*

- About.com: Saving Energy: What Are Compact Fluorescent Bulbs and How Do They Save Energy: *http://saveenergy.about.com/od/efficientlighting/a/CFL.htm?iam=sherlock_abc*

- *An Inconvenient Truth* (DVD website and resources): *www.climatecrisis.net*

- Carbon Footprint Ltd.: *www.carbonfootprint.com/index.html*

- Centre for Science Stories: Earth Sciences: The Realization of Global Warming: *http://science-stories.org/category/subject/earth-sciences*

- Chris Jordan, Photographic Arts: Running the Numbers: An American Self-Portrait (interactive art displays the statistics of American life; many relate to environmental issues; see, for example of the art "works," *Light Bulbs* (2008), which depicts 320,000 lightbulbs, equal to the number of kilowatt-hours

of electricity wasted in the United States every minute from inefficient residential electricity usage via a zoom function): *www.chrisjordan.com/gallery/rtn/#light-bulbs*

- Conservation International: Climate Change: *www.conservation.org/learn/climate/Pages/overview.aspx*
- Disney Educational Productions: *Bill Nye the Science Guy: Electrical Current* ($29.99/26 min. DVD): *http://dep.disney.go.com*
- Energy Star: Compact Fluorescent Bulbs: *www.energystar.gov/index.cfm?c=cfls.pr_cfls*
- Environmental Defense Fund:

    Fight Global Warming: *www.fightglobalwarming.com/viewads.cfm*

    Make the Switch, Take the Pledge Campaign (cost-saving calculators and other resources): *www.edf.org/page.cfm?tagid=608*

- Global Warming and the Greenhouse Effect (various sources):

    www.climatechange.net

    http://zebu.uoregon.edu/1998/es202/l13.html

    http://en.wikipedia.org/wiki/Greenhouse_effect

    www.ucsusa.org/global_warming/science/global-warming-faq.html

    http://lwf.ncdc.noaa.gov/oa/climate/globalwarming.html

- How Stuff Works: How much coal is required to run a 100-watt lightbulb for a year? *http://science.howstuffworks.com/question481.htm*
- Illinois Mathematics and Science Academy Center for Problem-Based Learning: PBL resources: *http://pbln.imsa.edu/index.html*
- Lawrence Berkeley National Laboratory: Standby power: *http://standby.lbl.gov*
- National Association of Biology Teachers Position Statements: *www.nabt.org/websites/institution/index.php?p=35*

    Sustainability in Life Science Teaching

    Teaching About Environmental Issues

- National Center for Case Study Teaching in Science: The Petition: A Global Warming Case: *www.sciencecases.org/petition/petition.asp*
- National Energy Education Development (NEED) Project: *www.need.org*

Making Sense by Spending Dollars

- National Public Radio: All Things Considered:

    CFL Bulbs Have One Hitch: Toxic Mercury: *www.npr.org/templates/story/story.php?storyId=7431198*

    Global Warming: It's All About Carbon: NPR, five-part, online cartoon series: *www.npr.org/templates/story/story.php?storyId=9943298*

- National Resource Council: America's Climate Choices (reports include *Understanding and Responding to Climate Change*): *http://americasclimatechoices.org*

- National Science Teachers Association Position Statements:

    Environmental Education: *www.nsta.org/about/positions/environmental.aspx*

    Teaching Science and Technology in the Context of Societal and Personal Issues: *www.nsta.org/about/positions.societalpersonalissues.aspx*

- Nature Conservancy: Climate Change: What's My Carbon Footprint? (Carbon Footprint Calculator): *www.nature.org/initiatives/climatechange/calculator/?src=f1*

- NSDL Wiki: Classic Articles on Global Warming, 1824–1995 With Interpretative Essays: *http://wiki.nsdl.org/index.php/PALE:ClassicArticles/GlobalWarming*

- PhET Interactive Simulations: The Greenhouse Effect: *http://phet.colorado.edu/simulations/sims.php?sim=The_Greenhouse_Effect*

- Science Channel: Deconstructed "How does this device work?" video clips: *http://science.discovery.com/videos/deconstructed*

    See: How Incandescent Light Bulbs Work (2:50 min.) and Compact Fluorescent Bulb (2:50 min.)

- Union of Concerned Scientists: Citizens and Scientists for Environmental Solutions: Global Warming: *www.ucsusa.org/global_warming*

- U.S. Department of Energy (efficiency): *www.energy.gov/energyefficiency/index.htm*

    CFLs: *www.energysavers.gov/your_home/lighting_daylighting/index.cfm/mytopic=12050*

- U.S. Environmental Protection Agency:

    CFLs and Mercury: *www.epa.gov/mercury*

    Climate Change—Greenhouse Gas Emissions (links to interactive calculators): *www.epa.gov/climatechange/emissions/individual.html*

- U.S. Green Building Council: *www.usgbc.org*. See the Leadership in Energy and Environmental Design (LEED) green building certification program.

- Wikipedia: *http://en.wikipedia.org/wiki*. Search for: compact fluorescent lamp (CFL), fluorescent, incandescent, LED, and life cycle assessment (cradle-to-grave analysis)

Answers to Questions in Procedure, steps #1–#9

1. To compare the two types of lightbulbs fairly, one needs to consider the initial purchase price, the life of the bulb, the energy the bulb uses for a given quantity of light output, and the cost of electricity.

2. No question was included.

3. Packages sometimes confuse light output (lumens) with power demand (watts, a unit of energy use per unit of time) and energy (kilowatt-hours). These three variables are proportional to each other within a given type of bulb, but given the difference in the efficiencies of energy conversions they can be quite different across different types of bulb. In particular, CFLs are much more efficient at converting electricity into light than incandescent lightbulbs, which could perhaps more accurately be called "heat" bulbs.

4.

Bulb Type (Equivalent Light Output)	Cost of Bulb	Bulb Life (Advertised on Package)	Equivalent Cost	Cost of Electricity Over the Projected Life of 1 CFL (or 8 Incandescent Bulbs)
Incandescent (60 W)	$1	750 hrs. (x)	$8	0.06 kW × 6,000 hrs. × $0.10/kWh = $36
Fluorescent (15 W)	$3	6,000 hrs. (or 8x)	$3	0.015 kW × 6,000 hrs. × $0.10/kWh = $9

Making Sense by Spending Dollars

Total savings of using compact fluorescent = Cost of bulb(s) + Cost of Electricity = [$8 − $3] + [$36 − $9] = $32 over life of bulb

Note: These calculations are representative and would need to be adjusted depending on the local cost of lightbulbs and electricity and the advertised life of the bulbs.

5. Less electricity consumed means less fossil fuel had to be mined, with less pollution produced as a byproduct of the mining and subsequent burning of the fossil fuel in a power plant. One could also compare the relative mass of one versus eight bulbs and the environmental costs associated with producing and disposing of the equivalent number of bulbs. Also, incandescent lightbulbs produce significantly more heat than the compact fluorescent bulbs, which could lead to somewhat higher summer air conditioning bills, and somewhat lower winter heating bills depending on the location. In the latter case, it should be noted that heating one's house with lightbulbs is not nearly as efficient as heating one's house by burning a fossil fuel onsite, with no loss of energy related to transmitting electricity over wires from the power plant to the home use.

6.

Bulb Type (Equivalent Light Output)	Coal Produced to Power the Lightbulbs Over the Projected Life of 1 CFL (or 8 Incandescent Bulbs)
Incandescent (60 W)	0.06 kW × 6,000 hrs. × 8 lb. of coal/10 kW = 288 lb. of coal
Fluorescent (15 W)	0.015 kW × 6,000 hrs. × 8 lb. of coal/10 kW = 72 lb. of coal

Amount of coal "saved" (not burned) = 288 − 72 = 216 lb. of coal
Amount of carbon dioxide not produced as a result = 216 lb. of coal × (29 lb. CO_2/8 lb. "carbon" coal) = 783 lb. of carbon dioxide NOT emitted into atmosphere as a greenhouse gas over the life of one compact fluorescent lightbulb (CFL).

One person's informed action can make a difference, and multiplying this savings over millions of U.S. households is significant! According to the Environmental Protection Agency, "If every household in the U.S. replaced one light bulb with an ENERGY STAR qualified compact flueorescent light bulb (CFL), it would prevent enough pollution to equal removing one million cars from the road" (*www.enerystar.gov*). It would also save enough energy

to light three million homes per year! What a bright idea! (*Note:* Americans produce approximately 20 tons of CO_2 per capita per year.)

7. The relative scarcity of resources, geographic location of resources (i.e., do they need to be imported from a friendly foreign nation or a foe?), and mining and disposing or recycling costs for the materials used would need to be considered. Also, standard CFLs are not recommended for certain types of applications, such as those that use dimmers, timers, three-way fixtures, motion detectors, photocells, or clip-on shades or that are directly exposed to outside weather. Newer, specially designed CFLs are being marketed for these applications. Finally, CFLs, like all fluorescent bulbs, do contain trace amounts of mercury (approximately 5 mg/lamp) that pose a health hazard when the bulb is broken. Accordingly, CFLs should be recycled versus disposed of, as is the practice in the European Union (see Internet Connections: Wikipedia: life cycle assessment). However, even without mandated recycling, the EPA estimates that the extra coal burned to power incandescent lightbulbs releases even more mercury into the environment! Nonetheless, the presence of small amounts of mercury in CFLs does raise the issue of tradeoffs and the need for cradle-to-grave product management. Some municipal waste departments and large retailers in the United States already recycle old CFLs in their separate hazardous-waste collection sites. The European Union has decided that CFLs make the most sense and actually banned the future production of incandescent lightbulbs as of September 1, 2009. When the current stockpiles of incandescent lightbulbs are sold, stores will only sell energy-efficient lightbulbs.

8. You could check to see if the claims about relative average life spans of bulbs are accurate, or perhaps measure the relative amount of light versus "waste" heat produced by the two bulbs. CFLs operate at temperatures less than 100°F, versus halogen floor lamps that can reach 1,000°F.

9. Energy- and money-saving windows, water heaters, air ducts, attic insulation, higher-mileage hybrid cars, Energy Star appliances, water-reducing faucets, radiant floor heating, and the mantra "reduce, reuse, recycle" make both environmental and economic sense.

Section 5: Assessment to Inform Learning and Transform Teaching

Activity 21

A Terrible Test That Teaches
Curriculum-Embedded Assessment

Expected Outcome

A "terrible test" appears to be impossible to pass. However, it is intentionally designed to include common errors in item construction that, if noticed, allow the test-wise learner to ace the test without any prior knowledge of the fictitious content.

Science Concepts

Scientists collect data, search for repeating patterns in nature, and make and test informed inferences that seek to discover underlying explanatory causes and predictive implications of observed patterns. The design and analysis of teacher-created assessments require similar processes if they are to provide feedback to support student learning and inform teaching.

Science Education Concepts

This test is an unusual, discrepant-event assessment that invites teachers to think critically about the multiple purposes and design features of tests and assessment systems that measure what is most valued. They can use their conclusions to inform and improve curriculum and instruction. Tests that teach both the students and the teacher are psychologically rewarding in that they reinforce and extend learning as a meaning-making (rather than memorize-and-regurgitate) activity. Teachers can discuss how their experience taking this test is a *participatory analogy* for the confusion, frustration, and anxiety students experience when taking tests that don't seem to make sense or to be relevant to their interests. Designing effective assessments and making data-driven decisions based on the results are challenging professional responsibilities. Similarly, students can be taught a sense of disciplined, minds-on playfulness about preparing for, taking, and reviewing their test results.

Teaching a given unit of study can be viewed as an experiment where the teacher: (1) has certain assumptions or theories about students' prior understandings (that can be tested via *diagnostic* assessments), (2) makes informed, best guess, curriculum design plans that bridge the gap between students' preinstructional conceptions and correct scientific explanations, (3) implements an instructional plan with curriculum-embedded *formative* assessments to provide "just-in-time" feedback to students and (4) modifies ongoing curriculum and instruction based on the assessment results. As such, a quality assessment system involves much more than paper-and-pencil, end-of-unit, *summative* "tests" for grading purposes.

A Terrible Test That Teaches

Materials
Paper copy of the 20-item test (without the correct answers) for each learner

> **Points to Ponder**
>
> *Measure what is measurable and make measurable what is not.*
> —Galileo Galilei, Italian astronomer and physicist (1564–1642)
>
> *If you can measure that of which you speak and can express it by a number, you know something of a subject, but if you cannot measure it, your knowledge is meager and unsatisfactory.*
> —William Thomson/Lord Kelvin, Scottish mathematician and physicist (1824–1907)
>
> *Not everything that counts, can be counted and not everything that can be counted, counts.*
> —Albert Einstein, German American physicist (1879–1955)
>
> *After the sale of over one million copies of the Taxonomy of Educational Objectives—Cognitive Domain (Bloom et al. 1956) and over a quarter of century of use of this domain in preservice and in-service teacher training, it is estimated that over 90% of test questions that U.S. public school students are now expected to answer deal with little more than information. Our instructional material, and our testing methods, rarely rise above the lowest category of the Taxonomy—knowledge.*
> —Benjamin Bloom (1984), American educational psychologist (1913–1999)

Procedure

The following test for teachers (modified from unknown sources) is best introduced with a degree of seriousness that typically accompanies traditional, graded, high-stakes tests: "Clear off your desks, take out one piece of lined paper and a number-two pencil, and number 1 through 17 straight down the page. All 20 items are worth 5 points each, including the three-short answer questions, #18–#20. All eyes should stay on your own papers and no talking is permitted after the papers are distributed. Put forth your individual best effort."

Part I: Multiple-Choice Items

_____ 1. The purpose of the cluss in furmpaling is to remove
    A. cluss-prag   C. cloughs
    B. treamlis   D. plumots

_____ 2. Trassig is true when
    A. lusp trasses the vom
    B. the viskal flans, if the viskal is donwil or zortil
    C. the bolgo frulls
    D. dissles lisk easily

_____ 3. The sigla frequently overfesks the treslum because
    A. all siglas are mellious
    B. siglas are always votial
    C. the treslum is usually tarious
    D. no tresla are feskable

_____ 4. The fribbled breg will minter best with a(n)
    A. derst   C. sorter
    B. morst   D. ignu

_____ 5. Among the reasons for tristal doss are
    A. the sabs foped and the foths tinzed
    B. the kredges roted with the orots
    C. few rakobs were accepted in sluth
    D. most of the polats were thonced

_____ 6. Which of the following is (are) always present when trossels are being gruven?
    A. vost and rint
    B. vost
    C. shum and vost
    D. vost and plume

A Terrible Test That Teaches

_____ 7. The mintering function of the ignu is most effectively carried out in connection with
A. a razma tol
B. the groshing stantol
C. the fribbled breg
D. a frally sush

_____ 8. Which of the following is NOT an example of a bliptine?
A. zerkose   C. sixtose
B. plixtose   D. zepcone

_____ 9. Cluss-prags are a problem in
A. furmpaling   C. zapaling
B. berfaling   D. ringaling

_____ 10. A.   C.
B.   D.

Part II: True/False Items

_____ 11. Every bogel has only one quixal.

_____ 12. Usually derst are found with sorter.

Part III: Fill in the Blank

13. What do we call the tree that grows from acorns? _____

14. What do we call a funny story? _____

15. What do we call the sound made by a frog? _____

16. What do we call a golfer's swing? _____

17. What do we call the white of an egg? _____

Part IV: Short Answer

18. List two words that describe how you feel in general about taking tests.

19. List two words that describe how you feel about this particular test.

20. What do you feel are the purposes of this particular test?

Debriefing

When Working With Teachers

This test is designed to initiate critical, ongoing discussions (and related action research) about a number of issues related to assessment. The following are examples of discussion prompts:

1. What are the three types (or purposes) of assessment, and how and when should they be used across a given unit of study? What does the phrase "schools overtest, but underassess" mean?

2. How could this particular test be used for diagnostic, formative, and summative purposes?

3. What are the advantages and limitations of high-stakes paper-and-pencil tests? Consider especially the affective response of students who are not especially "test-wise" and suffer from test anxiety.

4. Given that conventional paper-and-pencil tests are mandated by external groups (e.g., the federal No Child Left Behind (NCLB) law, state education departments, college admissions), how can teachers design high-quality, instructionally sensitive tests (as part of a broader, multidimensional assessment system) and teach students how to take externally mandated and designed standardized tests?

5. Discuss the contrasting quotes from the three prominent physicists (Galilei, Lord Kelvin, and Einstein) concerning the issue of measurement in science and their relevance to science assessment in schools.

6. Bloom's quote can be used as an entry into exploring topics such as Bloom's taxonomy, multiple-choice tests, table of specifications, national and international assessments, and alternative and laboratory or performance assessments as part of an ongoing professional development study group (see Internet Connections and the resource books listed on pages 269–270).

A Terrible Test That Teaches

When Working With Students

By middle school, students are increasingly expected to demonstrate their knowledge on paper-and-pencil tests. Teachers can help students develop test-taking skills (e.g., awareness of built-in clues, the importance of carefully reading the complete items, and the affective orientation of disciplined playfulness that enables them to avoid stress-induced reactions of "brain freeze" or cognitive downshifting). Each test returned to students is an opportunity to both reteach certain concepts and teach students how to prepare for and successfully take tests. The quotes are also useful for getting students to think about the critical role of measurement in science. When working with middle school students, rather than having them individually take the complete test in one sitting, the teacher may want to select a subset of items for whole-class practice and analysis.

Resource Books on Assessment in Science

- Doran, R., F. Chan, P. Tamir, and C. Lenhardt. 2002. *Science educator's guide to laboratory assessment.* Arlington, VA: NSTA Press. This book provides a good overview, with examples and scoring rubrics that emphasize laboratory performance tasks and alternative assessments such as concept mapping, V-heuristic, and oral presentations.
- Enger, S. K., and R. E. Yager. 2001. *Assessing student understanding in science: A standards-based K–12 handbook.* Thousand Oaks, CA: Corwin Press. This book contains student and teacher surveys, questionnaires, observation checklists, and scoring rubrics that cut across six domains of assessment (i.e., concepts, processes, applications, attitudes, creativity, and NOS/world views).
- Keeley, P., F. Eberle, and L. Farrin. 2005. *Uncovering student ideas in science, vol. 1: 25 formative assessment probes.* Arlington, VA: NSTA Press. See also other books in this series that include easy-to-use activities to assess K–12 students' preconceptions.
- Liu, X. 2010. *Essentials of science classroom assessment.* Thousand Oaks, CA: SAGE Publications, Inc. This book is a practical overview of assessing preconceptions (diagnostic), ongoing learning (formative), and summative outcomes.
- Mintzes, J. J., J. Wandersee, and J. D. Novak, eds. 2000. *Assessing science understanding: A human constructivist view.* San Diego, CA:

Academic Press. This book is a research-based follow-up to the authors' book *Teaching Science for Understanding*. Both volumes are scholarly but with clear practical implications for the classroom teacher.

- National Research Council (NRC). 1996. *National science education standards.* Washington, DC: National Academies Press. *www.nap.edu/catalog.php?record_id=4962*. See especially Chapter 5, "Assessment in Science Education," pages 75–102.
- National Research Council (NRC). 2001. *Classroom assessment and the national science education standards.* Washington, DC: National Academies Press. This book promotes assessment systems aligned with NSES. Table 4.1 (p. 63) is a framework that includes more than 40 assessment methods or formats. *www.nap.edu/catalog.php?record_id=9847*
- National Research Council (NRC). 2001. *Knowing what students know: The science and design of educational assessment.* Washington, DC: National Academies Press. This is the most "research-heavy" book in this list. *www.nap.edu/catalog.php?record_id=10019*
- National Research Council (NRC). 2006. *Systems for state science assessment.* Washington, DC: National Academies Press. Provides a broad overview of state-level assessments but contains much food for thought for teachers as well. *www.nap.edu/catalog.php?record_id=11312*

Extensions

1. *More Terrible Text and Tests*: Students can be taught to be more effective readers and test takers by using syntactical clues and grammatical structures to "make some sense" of otherwise incomprehensible text (or gibberish). Unlike the sample questions on page 271, it is important that teacher-designed assessments stretch students to higher levels of Bloom's taxonomy (knowledge–comprehension → application, analysis, synthesis, and evaluation). The inclusion of alternative forms of assessment (e.g., performance-based) is also important.

A Terrible Test That Teaches

   a. *Jabberwocky Jebberish:* Excerpt from Lewis Carroll's poem "Jabberwocky":

   Twas Brillig, and the slithy toves—Did gyre and gimbel in the wabe—All mimsy were the borogoves—And the mame raths outgrabe.

   i. What were the toves doing? (gyring and gambling)
   ii. Where were they doing it? (in the wabe)

   b. *The Montillation of Traxoline* (widely distributed on the Internet; attributed to Dr. Judith Lanier, professor and dean emeritus, Department of Education, Michigan State University)

   It is very important that you learn about traxoline. Traxoline is a new form of zionter. It is montilled in Ceristanna. The Ceristannians gristerlate large amounts of fevon and then bracter it to quasel traxoline. Traxoline may be be one of our most lukized snezlaus in the future because of our zionter lescelidge.

   i. What is traxoline? (a new form of zionter)
   ii. Where is traxoline montilled? (in Ceristanna)
   iii. How is traxoline quaselled? (gristerlate large amounts of fevon and then bracter it)
   iv. Why is it important to know about traxoline? (It may be be one of our most lukized snezlaus in the future because of our zionter lescelidge.)

   Teachers and students alike will enjoy the following spoof of overly "scientific-sounding" explanations. This *Turbo Encabulator* video clip (1:49 min.) originated as an unrehearsed warm-up for a real Rockwell International automatic transmission video. Scientifically literate viewers will pick up that it is completely facetious: *http://biggeekdad.com/2010/11/turbo-encabulator*.

2. *No Child Left Behind and the Nature of "Knowing and Showing It":* When Working With Teachers (only). The Internet Connections include several humorous parodies of NCLB. Ask teachers to read the short essays and discuss the extent to which high-stakes NCLB and state-mandated, summative science assessments might inadvertently encourage poor "teach to the test, drill and kill" curricular

and instructional practices that lead to higher test scores but lower levels of student understanding and interest in science. The goal is to teach students to know and love science while simultaneously using a variety of assessments to support their conceptual growth and their ability to perform well both on mandated standardized tests and in life.

3. *Testing Teachers*: A variety of direct (e.g., observation rubrics) and indirect (e.g., student attitude surveys) teacher performance metrics are available for teachers to use in self- or critical friend and peer assessments (e.g., see the CLES, CSSS, MOSART, and RTOP websites in Internet Connections). The results of such formative assessments can be used to help identify areas for improvements in teachers' science content and pedagogical content knowledge and skills. See also Appendix B.

Internet Connections

- Alternative Strategies for Science Teaching and Assessment: links to 20 types of alternatives: *http://science.uniserve.edu.au/school/support/strategy.html*
- American Chemical Society Division of Chemical Education Examinations Institute:

    High school paper-and-pencil tests and laboratory assessments: *http://chemexams.chem.iastate.edu/materials/exams.cfm*

- *American Educator*, Summer 2007, "Can Critical Thinking Be Taught": *www.aft.org/pubs-reports/american_educator/issues/summer07/index.htm*
- *American Educator*, Ask A Cognitive Scientist column: *www.danielwillingham.com*

    "Why Students Think They Understand—When They Don't": *www.aft.org/pubs-reports/american_educator/winter03-04/cognitive.html*

- Annenberg/CPB Media: Interactive Workshops (eight free videos on demand): *Assessment in Math and Science: What's the Point? www.learner.org/workshops/assessment*

A Terrible Test That Teaches

- Bioliteracy (Conceptual assessment resources, including a 30-item online multiple-choice test): *http://bioliteracy.colorado.edu*
- Bloom's Taxonomy: *http://faculty.washington.edu/krumme/guides/bloom1.html*
- Bloom's Taxonomy Revised: *http://projects.coe.uga.edu/epltt/index.php?title=Bloom%27s_Taxonomy*
- Buros Center for Testing: Instructional Resources: *Standards for teacher competence in educational assessment of students*: *www.unl.edu/buros*
- Constructivist Learning Environment Survey (CLES): 25-item Likert scale of student perceptions of how well constructivist learning principles are used in their science classroom (includes related research papers): *http://surveylearning.moodle.com/cles*
- Council of State Science Supervisors (CSSS):

    Assessment Tools for Professional Development: *www.csss-science.org/tools.shtml* (e.g., Constructivist Learning Environment Survey [CLES], View of the Nature of Science [VNOS], Teacher Pedagogical Philosophy Interview [TPPI], and Modified Assessment of Science Knowledge [MASK]).

    State and Territory Science Assessments Information: (Excel download): *www.csss-science.org/assess.shtml*

- Educational Testing Service: (AP, GRE, PRAXIS, PSAT, SAT, etc.): *www.ets.org*
- Ericae.net Clearinghouse on Assessment and Evaluation: *http://ericae.net*
- Fair Test. National Center for Fair and Open Testing: *www.fairtest.org*
- Foundation for Critical Thinking (see Thinker's Guide series): *www.criticalthinking.org*
- Learning Point Associates: *Multiple Dimensions of Assessment That Support Student Progress in Science and Mathematics*: A critical issue synopsis: *www.ncrel.org/sdrs/areas/issues/content/cntareas/science/sc700.htm#sciencestd*

- MOSART: Misconception Oriented Standards-based Assessment Resource for Teachers: *www.cfa.harvard.edu/smgphp/mosart/about_mosart.html*
- *Multiple Choice Tests: Composition, Assembly and Analysis* (online manual by Fred Gillette, Testing Center of San Francisco State University; including Table of Specifications): *www.sfsu.edu/~testing/MCTEST/mainframe.html*
- National Assessment of Educational Progress (NAEP): "The Nation's Report Card"—Science: NAEP is conducted every four years on representative samples of students in grades 4, 8, and 12: *http://nces.ed.gov/nationsreportcard/science*
- National Association for Research in Science Teaching: *Research Matters to the Science Teacher:* Definition and Assessment of the Higher-Order Cognitive Skills and Teaching Conceptual Understanding to Promote Students' Ability to Do Transfer Problems: *www.narst.org/publications/research.cfm*
- National Center for Research on Evaluation, Standards and Student Testing (CRESST; search for free downloads related to K–12 science assessments): *www.cse.ucla.edu*
- *National Science Education Standards* (NRC 1996): see Chapter 5, "Assessment," pp. 75–101: *www.nap.edu/catalog.php?record_id=4962*
- National Science Teachers Association Position Statement: Assessment: *www.nsta.org/about/positions/assessment.aspx*
- No Child Left Behind (NCLB) at the U.S. Department of Education: *www.ed.gov/nclb/landing.jhtml*
- Nature of Science (NOS) Assessments web links: See Internet Connections for Activity #7
- NCLB and Humorous Parodies:

    The Football Version: (Original author unknown; widely circulated on the web):

    www.hoagiesgifted.org/nclb_sports.htm

    www.janebluestein.com/articles/football.html

    http://edublog.teacherjay.net/2009/01/15/no-child-left-behind-football-version (expanded)

A Terrible Test That Teaches

No Cow Left Behind: *www.storm-lake.k12.ia.us/Admin/Pages/ncowlb.htm*

The Best Dentist: *http://grantpta.missouri.org/best_dentist.htm*

Track and Field Parable for NCLB: *http://mathforum.org/kb/thread.jspa?threadID=1954698&tstart=105*

- Performance Assessment Links in Science (PALS) (online standards-based tasks): *http://pals.sri.com*

- Programme for International Student Assessment (PISA), Organization for Economic Cooperation and Development (OECD): assesses science literacy of students at the end of compulsory education (~ age 15) on a three-year cycle. PISA is heavy on real-world applications: *www.pisa.oecd.org* and *http://nces.ed.gov/surveys/pisa*. See also the October 2009 special issue of the *Journal of Research in Science Teaching* 46 (8): Scientific Literacy and Contexts in PISA Science: *http://www3.interscience.wiley.com/journal/122604861/issue*.

- Reformed Teaching Observation Protocol (RTOP): A 25-item teacher performance checklist for research-informed instructional strategies. Online Training Workshop for Using RTOP: *http://physicsed.buffalostate.edu/AZTEC/RTOP/RTOP_full*

- Scientific Work Experience Programs for Teachers (SWEPT): Student Outcomes Study: Instruments: NSF-funded study includes pre-post student attitudes about science and science instruction and pre-post cognitive measures (in biology and chemistry) drawn from NAEP, TIMSS, and SAT, plus teacher attitude and practices surveys—the SWEPT instruments are all in the public domain: *www.sweptstudy.org/instruments.html*

- Table of Specifications: Insuring Accountability in Teacher Made Tests: article from *Journal of Instructional Psychology* 31 (June 2004): pp. 115–129: *http://findarticles.com/p/articles/mi_m0FCG/is_2_31/ai_n6130123*

- Trends in International Mathematics and Science Study (TIMSS): assesses 4th and 8th graders in more than 40 countries on a four-year cycle:

    IES National Center for Education Statistics: *http://nces.ed.gov/timss*

TIMSS and PIRLS International Study Center, Boston College: *http://timss.bc.edu*

- WestEd Partnership for the Assessment of Standards-based Science: PASS assessment take about 2 hours and 40 minutes and are designed for grades 5, 8, and 10 to meet the NCLB requirements: *www.wested.org/cs/we/view/serv/84*
- WikiPedia: *http://en.wikipedia.org/wiki*. Search: Benjamin Bloom, concept inventory (links to downloadable, diagnostic, research-validated inventories in physics, chemistry, astronomy, biology, and statistics), NAEP, PISA, scientific misconceptions, and TIMSS.

Answers to Test on Pages 266–267

1. A: Part of the stem is repeated in the answer, which also stands out from the others as a hyphenated, longer word.
2. B: the most nuanced, qualified, and lengthy choice; it "sticks out."
3. C: most nuanced, qualified answer; "usually" fits with "frequently" in the stem.
4. D: Grammatically, the correct answer should begin with a vowel.
5. A: The stem called for more than one reason, and choice A provides two.
6. B: stands out as the shortest option; also, a response pattern seems to be evident.
7. C: Items #4 and #7 are essentially the same question.
8. D: sticks out with a *one* ending versus the other three that end in *ose*.
9. A: Items #1 and #9 are essentially the same question.
10. B: There is no question, so the only logical way to answer is to look to see if there is a pattern in the previous answers, which there is: A, B, C, D.
11. False: very specific and exclusive statement that allows no exceptions.
12. True: Nuanced, qualified answers are more likely to be true.
13. oak
14. joke
15. croak
16. stroke
17. albumin; some test-takers will mindlessly respond with *yolk* as a response set after the sequence of four rhyming answers to questions #13–#16. A similar response set can be elicited by asking students to spell the word *spot* aloud and then immediately

asking them, "What do drivers do when they come up to a green light?" Many students will respond with the word *stop* rather than *go*.
18. Most teachers will express some level of frustration or anxiety about this test. Teachers who are especially "test-wise" may experience fun when they discover the pattern of answers and purpose of the test.
19. Varied responses
20. Most teachers will have a sense that it is designed in part to have them develop a sense of empathy for their students, who may experience similar anxieties and frustrations with science tests. Some teachers will also see that the test highlights issues related to the design and purposes of assessments.

Answers to Debriefing Questions

1. Three broad categories of assessment are *diagnostic* (preinstructional probes to activate and assess learners' preconceptions and misconceptions); *formative* (curriculum-embedded means to monitor ongoing conceptual change in the context of daily in-class instructional activities and homework); and *summative* (end-of-unit tests [or other authentic assessments] to rank the order of students and grade them relative to one another [norm-referenced] or an external standard [criterion-referenced]). Summative assessment tends to be given the most attention and time; research indicates that most teachers underassess with respect to diagnostic and formative assessment. Diagnostic assessments tend to be primarily used with special education students. Curriculum and instruction can be made more intelligible to all learners when the teacher assesses preinstruction conceptions because new learning builds on previous learning. A large body of research indicates that students have persistent and tenacious misconceptions on a range of science concepts that, if not addressed, restrict or subvert new learning (see Appendix A). Discrepant-event demonstrations presented in a predict-observe-explain format are one of a variety of ways to diagnostically interview a whole class about what they know (or believe to be so) about a given scientific concept. Short paper-and-pencil surveys (or nongraded "tests") are another option—see Activity #22. Teachers commonly use formative assessments in the form of spontaneous

questions they ask and planned worksheets or problem sets they assign. Formative feedback informs both teachers and students and is critical in learning; practice makes perfect only if learners receive specific and timely performance feedback.

2. This test has the look and feel of a standard, mixed format, paper-and-pencil, end-of-unit *summative* assessment. However, it actually is better used as either a diagnostic assessment to introduce teachers (or students) to the idea of skills to employ when designing (or taking) tests, or as a fun, curriculum-embedded, formative assessment to challenge learners after they have received some instruction on how to design (or how to take) tests.

3. Paper-and-pencil summative tests represent one way to assess a range of levels of understanding (i.e., Bloom's taxonomy) about science concepts. Care must be taken to have tests align with the relative time spent on different concepts during instruction (that are in turn aligned with the curriculum standards—see the Table of Specifications link in the Internet Connections). It is difficult to eliminate reading skills, test-taking skills, and confidence as factors that affect student outcomes. Accordingly, it is important to use a diversity of assessments that allow students to demonstrate their understanding in a variety of ways (i.e., multiple intelligences), formats (e.g., performance-based, authentic, project-based), and time frames (e.g., on-demand tests, daily homework, long-term projects).

4. Volumes have been written on the topic of paper-and-pencil assessments. Consider, for example, how carefully chosen distractors for multiple-choice items can be used to assess specific misconceptions (see the MOSART website in the Internet Connections) and how short answer and essay items can be used to teach students scientific argumentation skills.

Diagnostic Assessment
Discrepant Event or Essential Educational Experiment?

Expected Outcome

Learners are asked to complete sample diagnostic assessments that model how nongraded, preinstructional "tests" support learner-centered curriculum and instruction. These kinds of tests, the specific questions asked, and the misconceptions and conceptual holes that are elicited will likely be discrepant events for students and teachers alike. Although not conventional discrepant-event demonstrations by themselves, diagnostic assessments can be used in conjunction with inquiry-oriented demonstration-experiments and hands-on explorations.

Science Concepts

Carefully observing and, when possible, measuring pre-existing conditions in a system is a critical prerequisite for designing productive scientific experiments to assess change over time. As an analogy, consider how a medical doctor critically examines a patient's initial condition and completes a thorough diagnosis before recommending a therapeutic treatment or pharmaceutical prescription. Additional testing also often is done during the course of treatment to see how the treatment is working and to make changes as necessary. Two content domains, flat versus spherical Earth (scientific reasoning) and plants, are used here to show how to activate and assess misconceptions that are common and tenacious and that pose barriers to new learning. It is the nature of both science and learning to build on (and at times reconstruct) the foundation of earlier ideas, so identifying erroneous ones is essential to progress.

Science Education Concepts

Education is a profession that could benefit from better pretreatment diagnosis and midtreatment formative assessments and more attention to pre- and postinstruction summative gains. Teachers commonly use both informal and formal formative assessments (e.g., in-class activities, oral questions, low-stakes quizzes, and homework assignments) and teacher-designed and high-stakes, state-mandated summative-type tests. Diagnostic assessments provide critical information about learners' prior conceptions that point to solid foundations and potential stumbling blocks (i.e., misconceptions—see Appendix A) that subsequent curriculum and instruction can build on or deconstruct and reconstruct as necessary. In short, the assessments help teachers assess their students' zone of proximal development (ZPD), where appropriately targeted curriculum and instruction can support conceptual growth and development of higher-order thinking skills (i.e., Bloom's levels of application, analysis, and synthesis). Teachers may also discover that they themselves are not immune to harboring science misconceptions or that they have conceptual "holes" in their own understanding of science.

Appropriately designed diagnostic probes used in the Engage phase of the 5E Teaching Cycle (Bybee et al. 2006) can also activate

Diagnostic Assessment

attention and motivate a need-to-know interest in the specific content to be learned. Humorous titles, real-world practical applications, and an opportunity for students to express their interests all convey the message that the teacher cares about the students as much as they do about the content and that the classroom is an invitation to inquiry (i.e., "Students won't care how much you know until they know how much you care.").

The process of designing, implementing, evaluating, and revising integrated, "intelligent" curriculum (Where do we want to go?)-instruction (How shall we get there?)-assessment (Where are we at any particular point in the instructional sequence relative to our target destination?) is analogous to the experimental, iterative process of successive approximation that characterizes scientific inquiry. Published research on student alternative conceptions, lesson study groups, and individual teachers' assessment-informed action research in their own classrooms are all means to improve the connection between pedagogical interventions of teachers and improved learning outcomes of students. Pre- and postinstruction gain scores and gap analyses are especially effective means of evaluating progress with respect to tenacious misconceptions. Diagnostic assessments may also be targeted to measure students' evolving understanding of the nature of science (NOS) and their attitudes about science in general or school science in particular (see Internet Connections for examples).

Materials

Distribute individual copies of the model assessments on pages 284–286. If available, Scantron grading sheets are useful for rapid scoring and simple statistical analysis of results in grades 5–12 classrooms.

Points to Ponder

The liberty of choice [of scientific concepts and theories] is of a special kind; it is not in any way similar to the liberty of a writer of science fiction. Rather, it is similar to that of a man engaged in solving a well-designed word puzzle. He may, it is true, propose any word as the solution; but, there is only one word which really solves the puzzle in all its parts. It is a matter of faith that nature—as she is perceptible to our five senses—takes the character of a well-formulated puzzle. The successes reaped up to now by science ... give a certain encouragement to this faith.

—Albert Einstein (1879–1955), in "Physics and Reality," in *Ideas and Opinions*, pp. 290–323; originally published in the *Journal of the Franklin Institute* 221 (3).

What is taught, how it is taught, and how the success of that teaching is interpreted must all be part of a continuous process, with learning integrated within and across disciplines and between school and non-school performance settings. Assessment must be embedded in the teaching and learning process, rather than "delivered" out of context at discrete moments throughout a student's career. It should provide information about student potentials as well as progress, and be motivating to students, teachers, and schools as it illuminates compelling goals along with means for achieving them.

—*Learner-Centered Curriculum and Assessment for New York State* (NYSED, March 1994)

Diagnostic Assessment

Procedure

Teacher-learners complete one or both of the following diagnostic assessments as models for preinstructional tests that are relatively quick to administer, target common and tenacious misconceptions, provide instructors with some sense of the learners' prior experiences with the science concepts, and give learners a sense of the subsequent instructional unit's focus.

When using diagnostic assessments with grades 5–12 students at the start of a new instructional unit, teachers may want to introduce the Engage phase with one or more relevant discrepant-event activities in this book or the earlier *Brain-Powered Science* (2010) before asking students to complete the paper-and-pencil, nongraded "test." It is important that the teacher convey the message that diagnostic assessments are no-risk, nongraded surveys of students' prior knowledge and experiences, the results will be used to help the teacher better tailor the instruction to students' needs and interests, and students can benefit from getting a sneak preview of the focus of the unit. Regular use of diagnostic assessments can also lower students' anxiety about formal, graded assessments, both by increased experience with paper-and-pencil tests and by way of pre-exposure to some of the possible test items.

Sample Assessment #1

A Dialogue/Debate Between Dueling Theories: An Assessment of Scientific Reasoning

Historically, there was a lot of discussion and debate about whether the Earth is flat or spherical. Place the letter **F** in front of claims or hypotheses that *if true* would support a **F**lat Earth Theory. Place the letter **S** in front of claims or hypotheses that *if true* would support a **S**pherical Earth Theory. After filling in each blank line with F or S, draw a line through any claim that you feel can be or has been shown to be incorrect. On the back of this paper, write down any additional facts that you believe help support the more compelling "correct" theory.

1. ___ The Earth looks flat from a ground-based perspective and to people riding in airplanes.

2. ___ Ships that travel in a straight line eventually risk falling off the Earth's edge.

3. ___ If an airplane travels consistently in one direction for long enough, it will eventually return to its point of departure.

4. ___ Globes are physical models that approximately reflect the reality of our planet.

5. ___ People who live "down under" (in the Southern Hemisphere) risk falling off the Earth.

6. ___ A number of satellites are in a geocentric orbit around the Earth (e.g., GPS).

7. ___ The shadow of the Earth that sweeps across the full Moon during a lunar eclipse is curved (never elliptical or a straight line).

8. ___ A ship sailing toward the horizon gradually sinks from view, with the mast being the last thing seen; conversely, when the ship is returning, the mast is the first part to be seen.

9. ___ Pictures of the Earth from space show that it is spherical.

10. ___ Two-dimensional maps are physical models that reflect the reality of our planet.

11. ___ Gravity tends to make all large, celestial bodies (including Earth) roughly spherical.

12. ___ Traveling north or south from the equator, one sees different constellations at different latitudes.

13. ___ Answer the following question on the back of this page: If the spherical Moon rotates on its axis *and* revolves around the spherical Earth, how can you explain the observation that the "face" of a full Moon always looks the same (i.e., we never see the "dark side" of the Moon)?

Diagnostic Assessment

Sample Assessment #2

"Rooting" for Plants: So Much More Than Passive and Potted

Most people notice the unique nature and diversity of plant life within the animal kingdom more readily than they do when looking at the plant kingdom. In this unit, we are going to explore the fascinating "private lives" of plants. Please complete the following nongraded assessment to give me a better sense of what you already know about plants.

1. Plants are a lower or less-evolved form of life than animals. — True / False
2. A major characteristic of plants is that they do not move, whereas animals are always mobile. — True / False
3. All organisms that can photosynthesize are considered plants. — True / False
4. Plant food as purchased in garden stores supplies energy for the plant. — True / False
5. Carnivorous plants are not able to make enough food for themselves and therefore must consume insects as a supplemental food source. — True / False
6. Plants reproduce asexually, whereas animals reproduce sexually. — True / False
7. Some plants can live for several thousand years. — True / False
8. Throughout history, plants have been a primary source of food, medicines, clothing, building materials, dyes, and poisons. — True / False
9. Both plants and animals are multicellular organisms with membrane-bound cell organelles. — True / False
10. Plants undergo photosynthesis but do not perform respiration. — True / False
11. Plants are the source of food for most animals directly or indirectly. — True / False
12. Plants have evolved structural shapes and chemicals to entice some birds, bees, and bats to help them cross-pollinate or mate with other plants. — True / False
13. Some seeds can lie in a dormant state for thousands of years, then germinate when the environmental conditions are right. — True / False
14. Only animals can respond to their environments; plants are not able to respond to external stimuli. — True / False
15. Seeds are alive and show the characteristics of living organisms. — True / False
16. The search for plant products was a primary reason for European sea voyages of exploration and discovery. — True / False
17. Most of the nonwater weight of plants is due to the plant's ability to convert carbon dioxide from the air into carbohydrates. — True / False
18. Most of the nonwater weight of plants comes from minerals absorbed from the ground. — True / False

19. Plants and animals are interdependent; they need each other.   True   False
20. Mushrooms are a form of plant life.   True   False
21. All plants have roots that anchor them in soil and absorb minerals.   True   False
22. I've had experience hiking through natural areas where cars, roads, lights, and buildings cannot be seen or heard.   True   False
23. My house contains living plants and/or we grow fruits or vegetables in a home garden.   True   False
24. One question I have about plants is ...

Diagnostic Assessment

Debriefing

When Working With Teachers

After distributing or projecting the answers for teachers to quickly check their responses, focus the discussion on the advantages of using these kind of preinstruction, nongraded diagnostic assessments (i.e., rather than discussing any of the specific test items at length). If both tests are used, discuss the following differences in design:

Assessment #1 assesses the learner's ability to use empirical evidence, logical argument, and skeptical review to weigh the relative merits of the Flat Earth versus Spherical Earth theories. If desired, teachers and students can explore any of several websites about the Flat Earth Society (see Internet Connections). Also, the media literacy "documentary" film *In Search of the Edge: An Inquiry Into the Shape of the Earth* makes arguments for the Flat Earth Theory that are remarkably convincing to scientifically uninformed, noncritical viewers (see Internet Connections). Joy Hakim's book *The Story of Science: Aristotle Leads the Way* (2004) includes an extended discussion of ancient arguments for a round Earth (chapter 9, pp. 72–85, and chapter 18, pp. 160–165). Carl Sagan's 1980 *Cosmos* series on PBS recreated this same classic experiment of the Greek natural philosopher Eratosthenes (see Internet Connections).

Assessment #2 is a longer, forced-choice, true/false test that focuses on common misconceptions about plants. It also contains several items that ask students to note their prior experiences with plants. Most students tend to think of plants (approximately 260,000 known species) as relatively uninteresting, unimportant, fairly uniform in structure, semi-living objects that have some aesthetic appeal and are an optional part of our diet. Conversely, they "see" that biology is primarily about much more exciting, macroscopic, mobile, vertebrate animals (approximately 50,000 species) that are more like humans in their forms and functions (and, in the case of mammals, often appear "cute"). Our relative plant myopia and animal chauvinism are misconceptions that need to be challenged throughout the entire year rather than only in a chapter or two devoted to plants. Plants are essential components of the biosphere: They harness the energy from the Sun and convert it to forms that can be used by animals; generate the free oxygen needed for respiration by the

majority of living species; help regulate and purify the atmosphere; and provide food, shelter, clothing, and medicine for humans. Plants deserve as much curricular consideration and ecological preservation as animals. See Internet Connections: American Society of Plant Biologists.

These two sample diagnostic assessments represent a small sample of the wide variety of paper-and-pencil diagnostic probes. Interested teachers can explore the Internet Connections sites (in Activity #21) on concept inventories, MOSART, NAEP, PISA, scientific misconceptions, and TIMSS; research-based compilations of misconceptions (e.g., Driver et al. 1994; Duit 2009; Kind 2004); and sourcebooks (see Activity #21) for other test item formats (e.g., multiple choice and short answer) and interesting, misconception-based questions that could be used as possible test banks and models for their own assessments. If assessments are aligned to important curriculum standards and engage student interests, and the data collected are used to inform subsequent instruction, the phrase "teaching to the test" takes on a much more positive and proactive meaning than it commonly connotes.

If time permits, have teachers consider the Einstein quote with an eye to how it operationally defines the nature of science with respect to using both deductive logic and inductive experimentation to resolve scientific "puzzles." This quote and the one from New York State Education Department can also be used to help teachers reflect on the experimental nature of science teaching that uses an assessment system that includes curriculum-embedded diagnostic, formative, and summative assessments to synergistically link curriculum-instruction-assessment. Note that as the curriculum and instruction progress across the 5E Teaching Cycle of Engage, Explore, Explain, Elaborate, and Evaluate, the analogical roles of the teacher-assessor shift from diagnostician to coach or personal trainer (formative) and finally to judge and jury (summative).

Diagnostic Assessment

When Working With Students

In grades 5–12 classroom settings, teachers who regularly use diagnostic assessments may wish to use a Scantron grading machine to rapidly get a holistic picture of their students' prior knowledge. Diagnostic assessments that are used during the Engage phase of the 5E Teaching Cycle should not be reviewed with students, and answers should not be provided to students during this phase. Point out to the students that during the course of the unit, they should be able to correct their own answers, and you will also try to address any of the questions and/or interests they listed during the unit. You may wish to reuse a select subset of these same items on a graded postinstruction, end-of-unit test to look at gain score with respect to specific misconceptions. Though generating good assessment items is part of the intellectual challenge and responsibility of teaching, internet sites and books are great resources for teachers to draw on to avoid reinventing the wheel. See the reference books on creative problem-solving in science listed in Activity #11.

Extensions

1. *5 Easy Steps to an Intelligent Assessment System*: Teachers can explore how diagnostic, formative, and summative assessments can be creatively coordinated in a planned, curriculum-embedded fashion through the five phases of Engage, Explore, Explain, Elaborate, and Evaluate (Bybee et al. 2006). My next book, *Even More Brain-Powered Science*, will focus on modeling how to design 5E units.

2. *Assessing Attitudes About Science*: Assessing students' attitudes about science and science instruction is an important complement to content-focused assessments. Though students should never be graded on their attitudes, their attitudes do strongly influence their learning efforts and outcomes, as well as subsequent college and career choices. A quick Likert-scale-type survey of student attitudes at the start of a course and at periodic points throughout the year it can provide a teacher with a snapshot of the direction and extent of attitudinal shifts. See websites listed in Activity #21 (e.g., CLES, CSSS, and SWEPT), those in

Internet Connections (below), and links to the nature of science (NOS) instruments in Activity #7.

Internet Connections

See sites listed in Activity 21 and the following:

- American Society of Plant Biologists: *Principles of Plant Biology*: *www.aspb.org/education/foundation/principles.cfm*

- *Cosmos* (Carl Sagan's 1980 series; see episode #1, "The Shores of the Cosmic Ocean," for the segment on Eratosthenes's argument for a spherical Earth):

    www.hulu.com/cosmos

    http://video.google.com/videoplay?docid=4192793093886714469#docid=-7783461262084518299

- Disney Educational Productions: *Bill Nye the Science Guy: Flowers* and *Plants* ($29.99/26 min. DVD): *http://dep.disney.go.com*

- Flat Earth Society: *http://theflatearthsociety.org/cms*

    http://en.wikipedia.org/wiki/Flat_Earth_Society

    www.alaska.net/~clund/e_djublonskopf/FlatHome.htm

- In Search of the Edge: An Inquiry Into the Shape of the Earth (media literacy "documentary"): *www.bullfrogfilms.com/catalog/search.html*

- National Science Foundation (NSF), SRS Survey of Public Attitudes Toward and Understanding of Science and Technology (instrument + results): *www.nsf.gov/statistics/showsrvy.cfm?srvy_CatID=6&srvy_Seri=17*

- Physics Education Research Group, University of Maryland: Attitude Surveys (links): *www.physics.umd.edu/perg/tools/attsur.htm*

- Physics Education Research Group, University of Colorado: Colorado Learning About Science Survey: Physics, Chemistry and Biology versions: *www.colorado.edu/sei/class*

- *Relevance Of Science Education* (ROSE): International look at affective factors in learning: *www.uv.uio.no/ils/english/research/projects/rose*

Diagnostic Assessment

Answers to Assessment Questions

Sample Assessment #1

1. F The Earth looks flat from a ground-based perspective and to people riding in airplanes.
2. F ~~Ships that travel in a straight line eventually risk falling off the Earth's edge.~~
3. S If an airplane travels consistently in one direction for long enough, it will return to the point of departure *(by circumnavigating the spherical globe)*.
4. S Spherical globes are physical models that approximately reflect the reality of our planet.
5. F ~~People who live "down under" (in the Southern Hemisphere) risk falling off the Earth.~~
6. S A number of satellites are in a geocentric orbit around the Earth. *Such satellites are in a continual state of freefall around a spherical Earth.*
7. S The shadow of the Earth that sweeps across the full Moon during a lunar eclipse is curved (never elliptical or a straight line). *Note: Only one shape always casts a circular shadow—a sphere. This logical argument was known to Aristotle (384–322 BC). Later, Archimedes's friend Erathosthenes (276–196 BC) actually calculated the circumference of the Earth very close to the figure accepted today. Ancient Greek scholars did not believe in a flat Earth.*
8. S A ship sailing toward the horizon gradually sinks from view, with the mast being the last thing seen and, conversely, the mast is the first part to be seen when the ship is returning.
9. S Pictures of the Earth from space show that it is spherical.
10. F ~~Two-dimensional maps are physical models that reflect the reality of our planet.~~
11. S Gravity tends to make all large, celestial bodies (including Earth) roughly spherical.
12. S Travelling north or south from the equator, one sees different constellations at different latitudes.

13. If the Moon's rotation were *in sync with the Earth's rotation* as it revolved around the Earth, people on Earth would always see the same side of the Moon.

   Note: The Earth is actually not a perfect sphere, but rather bulges slightly at its equatorial "waist" due to its rotation about its axis.

Sample Assessment #2

1. False. Many students still think in terms of the outdated two-kingdom system that places all life into plant or animal categories. In the more modern five-kingdom classification system, cyanobacteria (formerly called blue-green algae) and algae are not considered plants or animals, but rather protists. (*Note:* The six-kingdom classification system splits the monera into two separate kingdoms, archaebacteria and eubacteria.) From an evolutionary perspective, modern plants arose after animals, but neither is considered more or less evolved or primitive. They have simply evolved different solutions for survival and reproduction. It is of course true that many animal species evolved more elaborate means of flexibly responding to the environment by way of their nervous and muscular systems.

2. False. Many plants have highly mobile seeds, some whole plants can change locations (e.g., resurrection plants and rootless water plants), and most plants exhibit geo- and phototropisms (i.e., directional growth is a type of movement on a slower scale). Conversely, some animals are "rooted" in one place for most of their lives (e.g., sea anemones and coral).

3. False. Prokaryotic cyanobacteria (previously called blue-green algae; kingdom *Monera*) and eurkaryotic algae (kingdom *Protista*) are not classified as plants even though both are photosynthetic autotrophs.

4. False. Autotrophic plants are multicellular eukaryotes that make their own food via photosynthesis. Mislabeled "plant foods" are actually fertilizers that commonly contain nitrogen, phosphorous, and/or potassium, which act as limiting elements for plant growth. The word *food* refers to organic compounds that are aerobically or anaerobically burned for energy.

5. False. As per the answer to #4, all plants make their own food. Carnivorous plants capture and consume small animals not for the energy content of their organic compounds, but for inorganic nutrients that are lacking in their environments (e.g., nitrogen).

6. False. Plants are commonly bisexual, without distinct morphologically different males and females, so people commonly (and incorrectly) think of them as using only asexual reproduction. Most plants have adaptations to prevent self-fertilization and support cross-fertilization with other plants of the same species. All plant seeds contain an embryo. It is true that most animals familiar to students reproduce sexually, with small, flagellated sperm fertilizing a larger, nonmotile egg.

7. True. For example, some species of coniferous trees (e.g., redwoods and sequoias) are among the Earth's tallest, heaviest, and oldest living organisms. One bristlecone pine tree called Methuselah is more than 4,838 years old. See *http://en.wikipedia.org/wiki/List_of_long-living_organisms*

8. True.

9. True.

10. False. Plants use respiration to break down glucose produced in photosynthesis to power their metabolism. They do not produce oxygen and food merely for the benefit of animals.

11. True.

12. True.

13. True. A record holder is a seed from the previously extinct Judean date palm that was coaxed to sprout after nearly 2,000 years of dormancy.

14. False. All living organisms must be able to respond to their environments to survive. In addition to the widespread geo- and phototropisms, some plants such as mimosa and various carnivorous plants can respond in a relatively rapid manner to certain touch-related stimuli.

15. True. However, seeds may lie dormant or seemingly lifeless for remarkably long periods of time.

16. True. However, the search for spices and dyes also was a major motivating force in European sea voyages.

17. True. Most people commonly (and incorrectly) assume that plant weight is largely due to absorption of minerals from the soil (as was true at the time of Jan Baptista van Helmont's willow tree experiment in the early 1600s). The conversion of carbon dioxide gas into glucose via photosynthesis is not commonly recognized as adding mass to the plant.

18. False. See #17.

19. True.

20. False. Fungi such as mushrooms were once grouped as plants in the old, outdated, two-kingdom system. All fungi are heterotrophs that are saprobes, parasites, or mutualistic symbionts (the latter, lichens, are photosynthetic due to their symbiotic relationship with algae).

21. False. Although most plants are terrestrial, some still live without roots floating on water and some are aerial. Also, more ancient plants such as bryophytes (e.g., mosses, liverworts, and hornworts) and early vascular plants (e.g., ferns) lack true roots.

22–24. Students' prior experiences in natural settings, as well their exposure to living plants in and immediately outside their homes, will vary both between schools and within classrooms. Similarly, their questions about and interest in plants are likely to vary widely. The hope is that by the end of a unit on plants, students will have a greater understanding of and appreciation for this essential and diverse form of life.

Appendix A

Alternative, Naive, Preinstructional, Prescientific, or Prior Conceptions Matter

Misconceptions, or a Rose by Any Other Name Is Still as Sweet (and/or as Thorny)

Despite the prevalence of the pedagogical misconception that equates teaching with telling and learning with listening, research has confirmed what effective teachers know about the learning-teaching dynamic. Learners are never passive blank slates to be written on or empty vessels to be filled by teachers (Bransford, Brown, and Cocking 1999; Donovan and Bransford 2005; Michael and Modell 2003). At any age, our brains actively (re-)construct our expanding sense of reality by creatively using the biological frameworks of our evolutionary heritage of genetically programmed but developmentally adaptable neural networks. Humans have an innate drive to make new synaptic connections (and strengthen and remake old ones) that compel us to seek out and learn from new experiences. The brain's neuroplasticity supports knowledge construction and reconstruction as an ongoing, lifelong process where new experiences are continually processed through the lens and build on the foundation of our unique, individual prior-knowledge bases.

Optimally, as our experiential bases grow, so do our interests, motivations, learning efforts, and understanding in a self-reinforcing, positive feedback loop (Dewey 1910, 1913, 1938). However, our brain's natural propensity for perceiving and/or conceiving patterns of explanatory relationships between experiences can also lead us to construct erroneous conceptions. Our abilities to make sense of both external sensory perceptions and internal conceptions is restricted by the biases and limitations that are built into our perceptual systems and concept-building neural networks (see *Brain-Powered Science*, Activities #4–#7 and #27–#29). These limitations are especially problematic in that many core scientific concepts are either beyond the reach of our unaided senses or are counterintuitive (see Activity #2 in this book; Cromer 1993; Wolpert 1992). Furthermore, the breadth, depth, and validity of our conceptions are based on the limited range of phenomena we have experienced and the external learning supports that were available to help us process those experiences. While expanding the latter is what effective curriculum-instruction-assessment is all about, even in the best learning environments it is inevitable that learners will draw some erroneous conclusions that will later need to be "unlearned" and replaced with better theories. It is the evolutionary nature of both science itself and individual learning to remain open to revision. Analyzing the sources, nature, and pedagogical implications of student errors is an important professional responsibility of both teachers and researchers (Chinn and Brewer 1998; Driver et al. 1994; Fisher and Lipson 1986).

Analogous to the production of movies, "miss-takes" in learning often are necessary steps along the path to an improved version of understanding. In movies (and TV shows), funny errors and other unplanned happenings sometimes become outtakes that are shown after the credits (or, in a few cases, actually make their way into the final cut as serendipitous, creative additions). Of course, unlike the movies, the ever-changing living brain never makes a final cut of its picture of reality. It is always potentially in a production and editing mode of reconstructing and expanding beyond past conceptions in light of new, ongoing experiences. Movie producers have the constraints of budget and time and the support of a team of internal reviewers and editors to provide formative feedback on their rough first takes or cuts.

Alternative, Naive, Preinstructional, Pre-scientific, or Prior Conceptions Matter

Similarly, individual students acting as their own mental movie producers face the constraints of scheduled summative assessments and unit changes and the support provided by interactions with their peers, teachers, textbooks, and nature (e.g., discrepant events). Any and all of these interactions can potentially provide scaffolded, "just-in-time," critical-friend feedback to students on the validity and generalizability of their current mental models and understandings.

Some of the erroneous ideas that students hold may be relatively innocuous and easy to replace in light of additional experiences that fill in conceptual "holes" or modify simple misunderstandings. However, nearly 50 years of research (e.g., Driver et al. 1994; Mintzes, Wandersee, and Novak 1998) has shown that many present-day science misconceptions are ubiquitous across ability, age, culture, educational background (i.e., science teachers themselves may ascribe to them), gender, and history (i.e., misconceptions may be similar to theories held by natural philosophers and scientists from earlier eras that have since been shown to be invalid); are robust, tenacious, and resistant to change (especially when they seem to work in everyday life) by conventional curricular and instructional approaches; may be undetected by conventional formative and summative assessments that overemphasize lower levels of Bloom's taxonomy (i.e., knowledge and comprehension) relative to higher-order thinking skills (i.e., application, analysis, synthesis, and evaluation); and represent barriers to understanding more scientifically valid concepts and conceptual networks of related ideas. A critical first step to help students recognize and reconstruct deeply held pre-existing, scientifically naive theories is for teachers to become aware of the following 10 potential sources of miseducation and pseudo-learning:

1. *Popular culture*: Much of what we learn comes from informal sources such as family and friends, media (e.g., movie special effects), children's literature, consumer advertisements, and pseudoscientific writings (see Activity #5). These sources typically have enough commonsense plausibility, situation-specific applicability, and source credibility for students to uncritically accept the validity of the information. Additional complications include the "foreign words for foreign things" nature of some scientific terminology and the loose use (or misuse) of scientific

terminology in everyday life. For example, terms such as *energy*, *work*, *force*, and *pressure* have restricted, distinct, nuanced, scientific definitions that are typically ignored and contradicted in everyday usage. Science teachers need to help students learn to reinterpret popular culture through the lens of critical reasoning and scientific skepticism.

2. *School culture and artifacts*: "Sins of omission and commission" in textbooks, by teachers, and on tests (i.e., what they omit and/or inappropriately include) speak with the voice of authority but may unwittingly generate or reinforce misconceptions about both science content and the nature of science. In some cases, efforts at simplification (e.g., visuals) or developmentally inappropriate curricula leave students with impoverished views of the truly "far-out" nature of reality. Examples include static, two-dimensional, off-scale, falsely colored drawings of nuclear atoms, cells, and our solar system that students are asked to believe on faith. For counterexamples related to the concept of atoms and molecules, see Activities #8 and #15 in this volume and Activity #20 in *Brain-Powered Science*. If a picture is worth a thousand words, visual representations, models, and analogies must be carefully selected and critically examined with an eye to helping students discover their limitations, lest they make erroneous cognitive connections. In addition to uncritically examined simplifications, research shows that teachers (and textbook authors) may actually share and inadvertently strengthen some of the very same misconceptions that students hold. Understandably, this is more common in newer, inexperienced elementary and middle school "science" teachers who did not major in science and in high school science teachers teaching a field of science outside their major or who are relatively new to teaching.

3. *Translation errors*: "Translation" errors can occur in both formal and informal instructional contexts. Individual learners use their unique conceptions of prior experiences and current motivations as filters to selectively screen their attention, focus their perceptions, and direct their (re)conceptualization of external reality (e.g., Activity #4). Thus, it is not unusual for there to be substantive differences between what is taught correctly by a teacher and what is actually caught by individual learners. These differences highlight

Alternative, Naive, Preinstructional, Pre-scientific, or Prior Conceptions Matter

the failure of a transmission- and passive-reception theory of teaching and learning and the power of a more learner-engaged, interactive, constructivist view, with its emphasis on continuous, curriculum-embedded assessment and "just in time" feedback (see *Brain-Powered Science*, Section III activities).

4. *Lack of prior experiences*: Another source of misunderstanding may be a lack of relevant prior experiences with the appropriate phenomenon and the mind's tendency to "abhor a vacuum" and fill in the gaps within a context of cognitive conservatism and intellectual inertia (e.g., Activity #9 in this book and Activities #25–#29 in *Brain-Powered Science*). We try to fill in gaps in our understandings, even if that requires forcing a round peg into a square hole. Discrepant-event activities are designed to shake up and overturn what we think we know by confronting us with results that are anomalous when viewed through the lens of our prior (mis)conceptions (e.g., Activity #14 in *Brain-Powered Science*). However, if only one such activity is used, it is easy for learners to see that event as a special case exception that might be applicable to school science and still hold on to their prior misconception in the real world.

5. *Everyday experiences*: Everyday experiences without the benefit of experimental controls and informed, carefully designed tests often seem to run counter to accepted scientific theories that have evolved over hundreds of years on an expanding foundation of empirical evidence, logical arguments, and skeptical review. For instance, friction causes objects set in motion to decelerate and stop and air resistance may cause different objects to fall at varying rates. Constant nonaccelerated, horizontal linear motion and the constant vertical acceleration of gravity as observed in artificial school science laboratory conditions seem to be the exceptions to the rule, rather than the rule (e.g., Activity #1). Interestingly, many misconceptions are similar to once accepted but now rejected, historically defunct "scientific" theories. Lamarckian evolutionary biology, alchemy-based chemistry, and Aristotelian physics all contain a certain amount of common sense, even though they have been proven incorrect.

Appendix A

6. *Placement in learning progression*: Developmental limits on formal conceptual reasoning ability and mathematical skills suggest that textbooks prematurely introduce and cover some science concepts without consideration of age-related readiness, experiential support, and conceptual scaffolding (e.g., Activities #13–#15 and #17). In these circumstances, it is inevitable that misconceptions will be created inadvertently. Research on and development of intra- and intergrade-level learning progressions are needed to design and refine field-tested K–12 curriculum sequences that enable teachers to help students make sense of the big ideas in science in ways that are developmentally appropriate (NRC 2007).

7. *Accessibility*: Built-in, species-specific sensory, neurological biases limit direct human access to only a small portion of the range and scales of external stimuli related to pressure- and heat-sensitive touch, frequencies, and intensities of reflected electromagnetic radiation/sight and sound/hearing and the types and concentrations of chemical tastes and odors/smells (see *Brain-Powered Science*, Section II). Additionally, we have difficulty conceptualizing the extreme range of exponential scales of time, matter, and space (e.g., powers of ten; see Activity #17). Instrumentation technologies, mathematical techniques, and the computational power of computers have greatly expanded the reach of our unaided senses and our sense of reality. The validity, reliability, and predictive and explanatory power of many modern scientific theories depend on empirical evidence gathered and analyzed with technologies that seem to be "black boxes" if they are not critically examined. For example, electron microscopes are an example of theory-embedded observations in that they allow us to "see" atoms by relying on electrons that are a subcomponent of the very atoms that are being pictured!

8. *Misunderstandings about science*: Misunderstandings about the "unnatural" nature of science and key conceptual ideas such as empirical evidence, experiment, necessarily naturalistic (and agnostic) methodologies and explanations, hypothesis, theory, law, and proof (versus falsifiability) make it difficult for students to understand the rules of the game (see Activity #2; Clough 2004). As a result, students may err in either of two extremes by becoming unduly skeptical and relativistic or uncritically

Alternative, Naive, Preinstructional, Pre-scientific, or Prior Conceptions Matter

accepting of scientific "facts." These extremes will most likely be elicited in instances that involve value-related matters such as science-technology-society (STS) issues or cases (e.g., Activities #18–20) of the supposed war between science and religion.

9. *Religious conflicts*: Some strongly held culturally disseminated religious beliefs run counter to certain scientific theories. Well-publicized examples include the anti-evolution, creationist views of biblical literalists and the anti-medical treatment stance of some religious sects. Other less obvious examples include our inability to sense the controlled molecular mania and atomic accounting principles that underlie the seeming stability in our own macroscopic appearances and personal identities and what happens to our physical components after death (e.g., compare Egyptian mummification to modern embalming and burial practices). Individuals who hold such beliefs typically are not aware of (or they fail to acknowledge) that the evidence for the scientific theories they reject cut across multiple, integrated scientific fields (e.g., evolutionary theory synergistically draws on astronomy, chemistry, geology and physics, not simply biology); and many religious traditions (including most Christian, Jewish, and Eastern faith traditions) do not sense an irreconcilable contradiction between their religious beliefs and science.

10. *Future research*: Although not recognized as misconceptions today, some of our current, provisionally accepted, scientifically valid theories will in fact be significantly modified and/or completely overturned in light of future scientific research. More experienced teachers may have even seen modifications in and the creation of new nomenclature, classification schemes, and theories showing up in the textbooks they have used throughout their careers. In biology, the endosymbiotic theory of eukaryotic cells (i.e., mitochondria and chloroplasts are believed to be descendants of free-living prokaryotes) and the new emerging science of epigenetics that suggests the possibility of an almost Lamarckian-type "evolution" are examples of the nature of scientific progress. In any case, science teachers should encourage students to critically question the answers of today's "best bet" science with the hope of inspiring them to contribute to tomorrow's groundbreaking discoveries. Science is not a closed book

of answers to be memorized and believed, but a way of knowing that is forever open to progressive improvement and extension.

Awareness of these possible sources and the nature of students' misconceptions provides teachers and curriculum developers with valuable information to help select appropriate discrepant-event activities and instructional sequences. However, as previously mentioned, simply presenting students with single anomalous instances that create cognitive conflict does not guarantee the desired conceptual change will occur in whole or even in part (Baddock and Bucat 2008). Students can perceive and reconceive a discrepant-event and anomalous data in a variety of ways (Cheek and O'Brien 1996; Chinn and Brewer 1998; O'Brien 2010):

1. They may fail to notice or "see" the discrepancy and will continue to incorrectly interpret related phenomenon inside and outside the science classroom through the lens of their flawed prior conceptions.

2. They may see the discrepancy but consciously choose to ignore it as either a noncompelling case or "magic trick"; interpret it as a special case of an in-school exception to the rule of their prior conceptions, which may otherwise continue to provide adequate explanations for real-world, out-of-school contexts; or create some blended version or mental amalgamation of their prior conceptions and the scientifically correct alternative (even if these are logically incompatible to an expert). Depending on the nature of the subsequent assessment tasks, how closely they match the anomalous event, and the ratio of new and old conception they retain, students' failure to make a substantive conceptual change may or may not be evident or seem particularly problematic to them or their teachers.

3. They may see and accept the validity of the discrepancy, become dissatisfied with their prior conception, and decide that a "new" scientific conception is more intelligible, plausible, and fruitful (Posner et al. 1982) in light of its broader generalizability and intellectual economy or parsimony. In this latter case, their old conception is essentially "overthrown and thrown out." Effective teachers use intelligent, integrated curriculum-instruction-assessment (CIA) (including a variety of differentiated

Alternative, Naive, Preinstructional, Pre-scientific, or Prior Conceptions Matter

instructional strategies) to increase the probability of this latter, most desired outcome occurring. They also serve as motivational coaches to get their students to intentionally commit to do the hard work of reconceptualization of their prior cognitive structures and reinterpretation of their prior experiences.

Effective conceptual-change-focused CIA activates and confronts deep-seated misconceptions and provides cognitive scaffolding that empowers learners to complete the challenging task of reconstructing "what they know that isn't so." In Piagetian terms, conceptual *accommodation* is a more demanding task that requires more time and support than *assimilation*-type learning, which is more a matter of filling in missing pieces to an otherwise well-visualized puzzle. Pedagogical planning to facilitate accommodation of the big and often outrageous ideas of science (e.g., Activity #2) ranges from the micro-level of individual teachers' classrooms to the macro-level of K–12 school systems and includes the following:

1. Regular use of *discrepant-event demonstration-experiments* that create cognitive conflict and challenge students to use what they know to predict-observe-explain (White and Gunstone 1992) and make adjustments and revisions in their understanding as warranted by empirical evidence, logical arguments, and skeptical review.

2. Regular, critical use of well-selected *analogies, models,* and *visual representations* that help students build bridges of understanding between things they are quite familiar with and understand and scientific concepts that seem foreign and mysterious (Camp and Clement 1994; Gilbert and Watt Ireton 2003; Hackney and Wandersee 2002; Harrison and Coll 2008; Hoagland and Dodson 1998; Lawson 1993).

3. A yearlong focus on lessons that teach students how to learn to read and read to learn science from textbooks, popular science journals, and other print and electronic media. *Even More Brain-Powered Science* (forthcoming) will include several activities and a variety of resources as idea starters for teachers who wish to focus more attention on this component of inquiry-based science instruction.

Appendix A

4. Within a given course or grade level, a logically sequenced 5E Teaching Cycle that provides multiple differentiated but focused experiences that afford students adequate time and instructional scaffolding (Bybee et al. 2006). Preinstructional diagnostic and curriculum-embedded formative assessments (i.e., tools such as concept mapping that make students' evolving conceptual networks and understanding visible) are especially important as sources of feedback to both teachers and the students themselves (see Activities #21 and #22). *Even More Brain-Powered Science* will include a number of model 5E Teaching Cycle activities.

5. Carefully articulated *learning progressions across grade levels* introduce concepts and increasingly more powerful scientific models at developmentally appropriate times and depths (Michaels, Shouse, and Schweingruber 2008; NRC 2007, 2010).

Collectively, these five concept-building approaches are analogous to selling the relative advantages of a new, improved car over an older, less-efficient one. The wise salesperson invites prospective buyers to carefully examine and test drive the new car to become more informed and "sold" on its features before asking the buyer to trade in their old car and invest in the new one. Constructivist curriculum-instruction-assessment approaches are the antithesis of authoritarian teachers and textbooks that "tell and test" passive students to "believe and regurgitate" something different than what seems to make sense to them.

Books, articles, and websites that describe specific, research-documented misconceptions (e.g., Driver et al. 1994; Duit 2009; Kind 2004; Meaningful Learning Research Group; Olenick 2008; Operation Physics 2010; search Google Scholar: concept + misconceptions) and assessment strategies give teachers the tools to help students progressively modify and eventually replace misconceptions with better informed, more scientifically valid conceptions. If time and energy are not devoted to this task, attempts to "layer" new instruction on top of old, inappropriate, scientifically flawed conceptions are analogous to putting a coat of fresh paint over a rusty surface. That is, in time, the new paint simply flakes off, once again exposing the rusting surface. The half-life of mindlessly memorized, misunderstood, "just the latest and greatest facts" types of science concepts may be quite short indeed. Building (and rebuilding as necessary) a solid conceptual foundation lays the groundwork for a lifetime of learning in science that learners can apply to predict, explain, and enjoy new phenomena both within and beyond the two covers of their textbooks and four walls of their classroom.

Appendix B

The S₂EE₂R* Demonstration Analysis Form

Science teachers, peer coaches, mentors, and supervisors can use this analysis form to analyze live or videotaped demonstration lessons to provide feedback to facilitate ongoing collegial conversations and collaborations to improve instructional effectiveness.

Demonstration Title	
Grade-Level Focus	
5E Phase (circle one)	Engage Explore Explain Elaborate Evaluate

Common Theme Focus:
    System
    Models
    Constancy and Change
    Scale	

Nature of Science Focus	
Science Content Focus	

This table should be filled in by the teacher-demonstrator prior to the demonstration.

* **S₂EE₂R** Criteria: **S**afe, **S**imple, **E**conomical, **E**njoyable, **E**ffective, and **R**elevant

O'Brien, T. 1991. The science and art of science demonstrations. *Journal of Chemical Education* 68 (11): 933–936.

O'Brien, T. 2010. *Brain-powered science: Teaching and learning with discrepant events*. Arlington, VA: NSTA Press. Appendix A, pp. 343–353.

Appendix B

Demonstration Title/Concept: _____

SCALE:
1 = Missing   2 = Unsatisfactory   3 = Basic/Improvement Needed
4 = Proficient   5 = Distinguished/Exemplary

Check [√] to highlight specific subcategories that warrant discussion and/or need improvement.

1. Evidence of Advance Preparation and Rehearsal

                                                               1 2 3 4 5

   _____ Clearly identified materials and projected image (and/or handout) of setup (as needed)

   _____ Operations that take time but need not be seen by students are done in advance.

   _____ Absence of last-minute scrambling; "staging" in place for lesson to flow

2. Appropriateness of Introduction in Setting Context

                                                               1 2 3 4 5

   _____ Clearly stated purpose (unless intentionally "hidden") and/or focus questions

   _____ Appropriate length and depth; sets stage and creates anticipation for FUNomenon

   _____ Minimal forecasting; lets the demonstration do the talking; wow and wonder before words

3. Visibility

                                                                1 2 3 4 5

   _____ Performed at an appropriate scale and elevation for all learners to see ("supersized")

   _____ Organized benchtop devoid of physical obstructions and visual distractions

   _____ Used suitable background and color and lighting for visual contrast

The S₂EE₂R* Demonstration Analysis Form

4. Verbal Audibility and Body Language Are Appropriate and In Sync
                                                                               1 2 3 4 5

   _____ Appropriate tone, volume, and pitch variation to support suspense and surprise

   _____ Enunciation, rate of speech, and use of pauses to create interest and engagement

   _____ Body language: facial expressions, hand gestures, and physical movement

5. Interest-Building, Motivational Aspects Activate Student Attention
                                                                               1 2 3 4 5

   _____ Attention-arousing, counterintuitive, discrepant quality creates enjoyment.

   _____ Enthusiasm, showmanship, storytelling, mystery, and/or humor captivate students.

   _____ Relevance to students' lives is clear.

6. Interactive, Participatory Style Catalyzes Cognitive Processing (NOT a "show and tell")
                                                                               1 2 3 4 5

   _____ Hands-on student assistants used if appropriate; minds-on engagement of all

   _____ Student↔FUNomenon + Student↔Student + Student↔Teacher Interactions

   _____ Quantity, quality, and pacing of teacher questions (i.e., wait time and Bloom's higher-order thinking skills)

7. Inquiry-Based, Investigative, Experimental Approach Models the Nature of Science
                                                                               1 2 3 4 5

   _____ Prediction, observation (data collection), and explanation (POE) analysis

   _____ Empirical evidence + logical argument + skeptical review frame discussion.

   _____ Invites multiple alternative hypotheses, questions, and ideas for further investigation

8. Safety, Disposal, and Economic Considerations

                                            1 2 3 4 5

    _____ Simplicity or complexity of means and materials used is appropriate (low cost/high benefit)

    _____ Physical and psychological safety for teacher and students (reward/gain > risk/pain)

    _____ Responsible, environmentally sound treatment of waste (if any)

9. Attainment of Purpose (It engages and educates rather than merely entertains.)

                                            1 2 3 4 5

    _____ Demonstration works as planned (*If not*, the "miss-take" is used as a teachable moment to model the nature of science.)

    _____ Draws out, builds on, and challenges pre- and misconceptions as necessary

    _____ Promotes conceptual, skill, and attitude growth; sets stage for future learning

10. Overall Organization and Execution (i.e., effectiveness and contribution to overall unit)

                                            1 2 3 4 5

    _____ Suitability as linked to placement in the overall unit (i.e., 5E phase)

    _____ Connects science concepts ("trees") to the common themes ("big picture forest")

    _____ Overall effectiveness

Appendix C

Science Content and Process Skills

BIOLOGICAL SCIENCES

Botany/Plants
22. Diagnostic Assessment: Discrepant Event or Essential Educational Experiment? (Test #2 Rooting for Plants)

Disease Transmission
18. Medical Metaphor Mixer: Modeling Infectious Diseases

Microscopy (Includes Powers of Ten and Scale Effects)
 8. Reading Between the Lines of the Daily Newspaper: Molecular Magic (Extension #1)

17. Metric Measurements, Magnitudes, and Mathematics: Connections Matter in Science (Extensions #1, #4, and #5)

18. Medical Metaphor Mixer: Modeling Infectious Diseases (Extensions #1–#3)

EARTH AND PHYSICAL SCIENCES

Acid-Base Chemistry (and Indicators)
12. Magical Signs of Science: "Basic Indicators" for Scientific Inquiry

Atomic and Molecular Theory
8. Reading Between the Lines of the Daily Newspaper: Molecular Magic
15. Measurements and Molecules Matter: Less Is More and Curriculum "Survival of the Fittest"
17. Metric Measurements, Magnitudes, and Mathematics: Connections Matter in Science (Extensions)

Density
3. Dual-Density Discrepancies: Ice Is Nice and Sugar Is Sweet
14. Archimedes, the Syracuse (Sicily) Scientist: Science Rules Balance and Bathtub Basics (Part 2 and Extensions)
15. Measurements and Molecules Matter: Less Is More and Curriculum "Survival of the Fittest" (Extension #1)

Earth Science and Geology
17. Metric Measurements, Magnitudes, and Mathematics: Connections Matter in Science (Extensions)
19. Cookie Mining: A Food-for-Thought Simulation
22. Diagnostic Assessment: Discrepant Event or Essential Educational Experiment? (Test #1 Dueling Theories: Flat Versus Spherical Earth)

Energy and Energy Conversion
1. Comeback Cans: Potentially Energize "You CAN Do" Science Attitudes
16. Bottle Band Basics: A Pitch for Sound Science

Science Content and Process Skills

Magnetism
10. Follow That Star: *National Science Education Standards* and True North

Physical and Chemical Changes
12. Magical Signs of Science

Recycling, Matter, and Energy Conservation Efforts
8. Reading Between the Lines of the Daily Newspaper: Molecular Magic
15. Measurements and Molecules Matter: Less Is More and Curriculum "Survival of the Fittest" (Extensions #2 and #3)
17. Metric Measurements, Magnitudes, and Mathematics: Connections Matter in Science (Extensions)
19. Cookie Mining: A Food-for-Thought Simulation
20. Making Sense by Spending Dollars: An "Enlightening" STS Exploration of CFLs, or How Many Lightbulbs Does It Take to Change the World?

Sound
16. Bottle Band Basics: A Pitch for Sound Science

Process and Reasoning Skills for Teaching-Learning Science

Measurement and Mathematical Skills
6. Scientific Reasoning: Inside, Outside, On, and Beyond the Box
13. Verifying Vexing Volumes: "Can Be as Easy as Pi" Mathematics
14. Archimedes, the Syracuse (Sicily) Scientist: Science Rules Balance and Bathtub Basics
15. Measurements and Molecules Matter: Less Is More and Curriculum "Survival of the Fittest"
17. Metric Measurements, Magnitudes, and Mathematics: Connections Matter in Science

Nature of Science (NOS)

The NOS as a way of knowing is a cross-disciplinary theme or "big idea" that runs through all 22 activities. The NOS is a special focus of the two activities in Section I and the seven activities in Section II. All of these activities (#1–#9) can be used in both life sciences (biology) and physical sciences (chemistry, Earth science, and physics) courses.

Pseudoscience and Critical Reading and Thinking

2. Unnatural Nature and Uncommon Sense of Science
5. Pseudoscience in the News: Preposterous Propositions and Media Mayhem Matters

Test-Taking Skills

21. A Terrible Test That Teaches
22. Diagnostic Assessment: Discrepant Event or Essential Educational Experiment?

Research Cited

Abd-El-Khalick, F., R. L. Bell, and N. G. Lederman. 1998. The nature of science and instructional practice: Making the unnatural natural. *Science Education* 82 (4): 417–436.

Aicken, F. 1991. *The nature of science.* 2nd ed. Portsmouth, NH: Heinemann.

American Association for the Advancement of Science (AAAS). 1993. *Benchmarks for science literacy.* New York: Oxford University Press. *http://project2061.aaas.org.*

American Psychological Association (APA). 1997. *Learner-centered psychological principles: A framework for school redesign and reform.* Washington, DC: APA Center for Psychology in Schools and Education. *www.apa.org/ed/cpse/LCPP.pdf.*

Association for Science Teacher Education (ASTE). 2004. ASTE position statement: Science teacher preparation and career-long development. *http://theaste.org/aboutus/AETSPosnStatemt1.htm.*

Baddock, M., and R. Bucat. 2008. Effectiveness of a classroom chemistry demonstration using the cognitive conflict strategy. *International Journal of Science Education* 30 (8): 1115–1128.

Banilower, E. R., S. E. Boyd, J. D. Pasley, and I. R. Weiss. 2006. *Lessons from a decade of mathematics and science reform: The local systemic change through teacher enhancement initiative.* Chapel Hill, NC: Horizon Research. *www.pdmathsci.net/findings/report/32.*

Bell, R. L. 2008. *Teaching the nature of science through process skills: Activities for grades 3–8.* Boston, MA: Pearson/Allyn and Bacon.

Bloom, B. 1984. The 2 sigma problem: The search for methods of group instruction as effective as one-to-one tutoring. *Educational Researcher* 13 (6): 4–16.

Bloom, B. S., M. D. Engelhart, E. J. Furst, W. H. Hill, and D. R. Krathwohl. 1956. *Taxonomy of educational objectives: Handbook I; Cognitive domain.* New York: David McKay Co. (For related resources see *http://faculty.washington.edu/krumme/guides/bloom1.html.*)

Bransford, J. D., A. L. Brown, and R. R. Cocking, eds. 1999a. *How people learn: Brain, mind, experience, and school.* Washington, DC: National Academies Press.

Bransford, J. D., A. L. Brown, and R. R. Cocking, eds. 1999b. *How people learn: Bridging research and practice.* Washington, DC: National Academies Press.

Bybee, R. W., ed. 2002. *Learning science and the science of learning.* Arlington, VA: NSTA Press.

Bybee, R. W., J. C. Carlson Powell, and L. W. Trowbridge. 2008. *Teaching secondary school science: Strategies for developing scientific literacy.* 9th ed. Upper Saddle River, NJ: Merrill Prentice Hall.

Research Cited

Bybee, R. W., J. A. Taylor, A. Gardner, P. Van Scotter, J. Carlson Powell, A. Westbrook, and N. Landes. 2006. *BSCS 5E instructional model: Origins, effectiveness and applications.* Colorado Springs, CO: BSCS. *www.bscs.org/pdf/5EFull Report.pdf* (65 pages) and *http://bscs.org/pdf/bscs5eexecsummary.pdf* (19 pages)

Camp, C. W., and J. J. Clement. 1994. *Preconceptions in mechanics: Lessons dealing with students' conceptual difficulties.* Dubuque, IA: Kendall/Hunt.

Chabris, C., and D. Simons. 2010. *The invisible gorilla: And other ways our intuitions deceive us.* New York: Crown Archetype. *www.theinvisiblegorilla.com/videos.html*

Cheek, D., and T. O'Brien. 1996. *New York Science, Technology and Society Education Project (NYSTEP) teacher guide.* Albany, NY: Research Foundation of the State University of New York.

Chiappetta, E. L., and T. R. Koballa Jr. 2010. *Science instruction in the middle and secondary school: Developing fundamental knowledge and skills for teachers.* 7th ed. Boston: Allyn and Bacon.

Chin, C., and J. Osborne. 2008. Students' questions: A potential resource for teaching and learning science. *Studies in Science Education* 44 (1): 1–39.

Chinn, C. A., and W. F. Brewer. 1993. The role of anomalous data in knowledge acquisition: A theoretical framework and implications for science instruction. *Review of Educational Research* 63 (1): 1–49.

Chinn, C. A., and W. F. Brewer. 1998. An empirical test of a taxonomy of responses to anomalous data in science. *Journal of Research in Science Teaching* 35 (6): 623–654.

Clark, R. L., M. P. Clough, and C. A. Berg. 2000. Modifying cookbook labs: A different way of teaching a standard laboratory engages students and promotes understanding. *The Science Teacher* 67 (7): 40–43.

Clough, M. P. 2002. Using the laboratory to enhance student learning. In *Learning science and the science of learning,* ed. R.W. Bybee, 85–94. Arlington, VA: NSTA Press.

Clough, M. P. 2004. The nature of science: Understanding how the game is played. In *The game of science education,* ed. J. Weld, 198–227. Boston, MA: Pearson/Allyn and Bacon.

Clough, R. L., and R. L. Clark. 1994a. Cookbooks and constructivism: A better approach to laboratory activities. *The Science Teacher* 61 (2): 34–37.

Clough, R. L., and R. L. Clark. 1994b. Creative constructivism: Challenge your students with authentic science experience. *The Science Teacher* 61 (7): 46–49.

Coalition of Essential Schools Northwest. N.d. Critical friends groups. *www.cesnorthwest.org/cfg.php.*

Cochran, K. F. 1997. Pedagogical content knowledge: Teacher's integration of subject matter, pedagogy, students, and learning environments. Brief. Research Matters to the Science Teacher. No. 9702. National Association for Research in Science Teaching. *www.narst.org/publications/research/pck.htm.*

Cocking, R. R., J. P. Mestre, and A. L. Brown, eds. 2000. New developments in the science of learning: Using research to help students learn science and mathematics. Special issue of *Journal of Applied Developmental Psychology* 21 (1): 1–135.

Colburn, A. 2004. Focusing labs on the nature of science: Laboratories can be structured to help students better understand the nature of science. *The Science Teacher* 71 (9): 32–35.

Council of State Science Supervisors (CSSS). Science and safety guides (free downloads). *www.csss-science.org/safety.shtml.*

Cromer, A. 1993. *Uncommon sense: The heretical nature of science.* New York: Oxford University Press.

DeBoer, G. E. 1991. *A history of ideas in science education: Implications for practice.* New York: Teachers College Press.

Dewey, J. 1910. *How we think.* Amherst, NY: Prometheus Books. (Available online at *www.brocku.ca/MeadProject/Dewey/Dewey_1910a/Dewey_1910_a.html*)

Dewey, J. 1913. *Interest and effort in education.* Boston, MA: Houghton Mifflin. *www.archive.org/details/interestandeffor00deweuoft.*

Dewey, J. 1916. *Democracy and education: An introduction to the philosophy of education.* New York: Macmillan. *www.ilt.columbia.edu/Publications/dewey.html.*

Dewey, J. 1938. *Experience and education.* New York: Macmillan Publishing.

Donovan, M. S., and J. D. Bransford, eds. 2005. *How students learn: Science in the classroom.* Washington, DC: National Academies Press.

Doran, R., F. Chan, P. Tamir, and C. Lenhardt. 2002. *Science educator's guide to laboratory assessment.* Arlington, VA: NSTA Press.

Driver, R., E. Guesne, and A. Tiberghein, eds. 1985. *Children's ideas in science.* Milton Keynes, UK: Open University Press.

Driver, R., A. Squires, P. Rushworth, and V. Wood-Robinson. 1994. *Making sense of secondary science: Research into children's ideas.* London: Routledge.

DuFour, R., and R. Eaker. 1998. *Professional learning communities at work: Best practices for enhancing student achievement.* Bloomington, IN: National Educational Service and Alexandria, VA: Association for Supervision and Curriculum Development.

Duit, R. 2009. Bibliography of students' and teachers' conceptions and science education. Kiel, Germany: Institute for Science Education, University of Kiel. *www.ipn.uni-kiel.de/aktuell/stcse/stcse.html.*

Enger, S. K., and R. E. Yager. 2001. *Assessing student understanding in science: A standards-based K–12 handbook.* Thousand Oaks, CA: Corwin Press.

Fensham, P., R. Gunstone, and R. White, eds. 1994. *The content of science: A constructivist approach to its teaching and learning.* London: Falmer Press.

Fisher, K. M., and J. L. Lipson. 1986. Twenty questions about student errors. *Journal of Research in Science Teaching* 23 (9): 783–803.

Gallagher, J. J. 2007. *Teaching science for understanding: A practical guide for middle and high school teachers.* Upper Saddle River, NJ: Merrill Prentice Hall.

Gardner, H. 1999. *The disciplined mind: What all students should understand.* New York: Simon & Schuster.

Garet, M. S., A. C. Porter, L. Desimone, B. F. Birman, and K. S. Yoon. 2001. What makes professional development effective? Results for a national sample of teachers. *American Educational Research Journal* 38 (4): 915–945.

Research Cited

Gilbert, S. W., and S. Watt Ireton. 2003. *Understanding models in Earth and space science*. Arlington, VA: NSTA Press.

Goleman, D. 1995. *Emotional intelligence: Why it can matter more than IQ*. New York: Bantam Books.

Good, R. G., J. D. Novak, and J. H. Wandersee, eds. 1990. Perspectives on concept mapping. *Journal of Research in Science Teaching, Special Issue* 27 (10). www.narst.org.

Grant, J. 2006. *Discarded science: Ideas that seemed good at the time*. Wisley, Surrey, UK: Facts, Figures & Fun.

Grun, B., and E. Simson. 2005. *The timetables of history: A horizontal linkage of people and events*. 4th ed. New York: Simon & Schuster/Touchstone Books.

Hackney, M. W., and J. H. Wandersee. 2002. *The power of analogy: Teaching biology with relevant classroom-tested activities*. Reston, VA: National Association of Biology Teachers.

Hagevik, R., W. Veal, E. M. Brownstein, E. Allen, C. Ezrailson, and J. Shane. 2010. Pedagogical content knowledge and the 2003 science teacher preparation standards for NCATE accreditation or state approval. *Journal of Science Teacher Education* 21 (1):7–12.

Harrison, A. G., and R. K. Coll, eds. 2008. *Using analogies in middle and secondary science classrooms: The FAR guide—An interesting way to teach with analogies*. Thousand Oaks, CA: Corwin Press.

Harvard-Smithsonian Center for Astrophysics, Science Education Department. MOSART: Misconception Oriented Standards-based Assessment Resource. *www.cfa.harvard.edu/sed/projects/mosart.html*. (See also *Minds of Our Own* and *A Private Universe*.)

Highet, G. 1950. *The art of teaching*. New York: Vintage Books/Random House.

Hilton, M., rapporteur. 2010. *Exploring the intersection of science education and 21st century skills: A workshop summary*. National Research Council, Board on Science Education. Washington, DC: National Academies Press. http://books.nap.edu/catalog.php?record_id=12771.

Hoagland, M., and B. Dodson. 1998. *The way life works: The science lover's illustrated guide to how life grows, develops, reproduces, and gets along*. New York: Three Rivers Press/Random House.

Hofstein, A., and V. N. Lunetta. 2003. The laboratory in science education: Foundations for the twenty-first century. *Science Education* 88 (1): 28–54.

Keeley, P., F. Eberle, and L. Farrin. 2005. *Uncovering student ideas in science, vol. 1: 25 formative assessment probes*. Arlington, VA: NSTA Press.

Keeley, P., F. Eberle, and J. Tugel. 2007. *Uncovering student ideas in science, vol. 2: 25 more formative assessment probes*. Arlington, VA: NSTA Press.

Kind, V. 2004. Beyond appearance: Students' misconceptions about basic chemical ideas. 2nd ed. *www.rsc.org/education/teachers/learnnet/pdf/LearnNet/rsc/miscon.pdf*.

Lawson, A. E., ed. 1993. The role of analogy in science and science teaching. *Journal of Research in Science Teaching, Special Issue* 30 (10). www.narst.org.

Lawson, A. E. 2010. *Teaching inquiry science in middle and secondary schools.* Thousand Oaks, CA: Sage Publications.

Lederman, N. G. 1992. Students' and teachers' conceptions of the nature of science: A review of the research. *Journal of Research in Science Teaching* 29 (4): 331–359.

Lederman, N. G. 1999. Teachers' understanding of the NOS and classroom practice: Factors that facilitate or impede the relationship. *Journal of Research in Science Teaching* 36 (8): 916–929.

Lederman, N. G., and M. L. Neiss. 1997. The nature of science: Naturally? *School Science and Mathematics* 97 (1): 1–2.

Liu, X. 2010. *Essentials of science classroom assessment.* Thousand Oaks, CA: Sage Publications.

Lortie, D. C. 1975. *Schoolteacher: A sociological study.* Chicago: University of Chicago Press.

Loucks-Horsley, S., P. W. Hewson, N. Love, and K. E. Stiles. 1998. *Designing professional development for teachers of science and mathematics.* Thousand Oaks, CA: Corwin Press.

Lunetta, V. N., A. Hofstein, and M. Clough. 2007. Learning and teaching in the school science laboratory. In *Handbook of research in science education*, ed. N. Lederman and S. Abell, 393–441. Mahwah, NJ: Lawrence Erlbaum.

Macknik, S. L., and S. Martinez-Conde. 2010. *Sleights of mind: What the neuroscience of magic reveals about our everyday deceptions.* New York: Henry Holt. *www.sleightsofmind.com*

Mayer, R. E. 2009. *Multimedia learning.* 2nd ed. New York: Cambridge University Press.

McComas, W. F. 1996. Myths of science: Reexamining what we know about the nature of science. *School Science and Mathematics* 96 (1): 10–16.

McComas, W. F., ed. 1998. *The nature of science in science education: Rationale and strategies.* Dordecht, The Netherlands: Kluwer Academic Publishers.

McComas, W. F., ed. 2004. The history and nature of science. *The Science Teacher, Special Issue* 71 (9).

McCombs, B. L., and J. S. Whisler. 1997. *The learner-centered classroom and school: Strategies for increasing student motivation and achievement.* San Francisco, CA: Jossey-Bass.

Meaningful Learning Research Group. n.d. Misconceptions conference proceedings. *http://www2.ucsc.edu/mlrg/mlrgarticles.html.*

Medina, J. 2008. *Brain rules: 12 principles of surviving and thriving at work, home and school.* Seattle, WA: Pear Press. Book, DVD, and website: *http://brainrules.net*

Michael, J. A., and H. I. Modell. 2003. *Active learning in secondary and college science classrooms: A working model for helping the learner to learn.* Mahwah, NJ: Lawrence Erlbaum.

Michaels, S., A. W. Shouse, and H. A Schweingruber. 2008. *Ready, set, science! Putting research to work in K–8 science classrooms.* Washington, DC: National Academies Press. *www.nap.edu/catalog.php?record_id=11882.*

Research Cited

Mintzes, J. J., J. H. Wandersee, and J. D. Novak, eds. 1998. *Teaching science for understanding: A human constructivist view.* New York: Academic Press.

Mintzes, J. J., J. H. Wandersee, and J. D. Novak, eds. 2000. *Assessing science understanding: A human constructivist view.* San Diego, CA: Academic Press.

Morowitz, H. J. 2002. *The emergence of everything: How the world became complex.* New York: Oxford University Press.

Mundry, S., and K. Stiles, eds. 2009. *Professional learning communities for science teaching: Lessons from research and practice.* Arlington, VA: NSTA Press.

National Academies of Sciences (NAS). 2007. *Rising above the gathering storm: Energizing and employing America for a brighter economic future.* Washington, DC: National Academies Press. www.nap.edu/catalog.php?record_id=11463.

National Commission on Excellence in Education. 1983. *A nation at risk: The imperative for education reform* (Stock No. 065-000-001772). Washington, DC: U.S. Government Printing Office. www.ed.gov/pubs/NatAtRisk/index.html.

National Commission on Mathematics and Science Teaching (NCMST) for the 21st Century. 2000. *Before it's too late.* Washington, DC: Department of Education. www.ed.gov/inits/Math/glenn/report.pdf.

National Commission on Teaching and America's Future (NCTAF). 1996. *What matters most: Teaching for America's future.* New York: Teachers College Press. www.nctaf.org/documents/WhatMattersMost.pdf.

National Commission on Teaching and America's Future (NCTAF). 1997. *Doing what matters most: Investing in quality teaching.* New York: Teachers College Press. www.nctaf.org/documents/DoingWhatMattersMost.pdf.

National Research Council (NRC). 1996. *National science education standards.* Washington, DC: National Academies Press. www.nap.edu/catalog.php?record_id=4962.

National Research Council (NRC). 2000. *Inquiry and the national science education standards: A guide for teaching and learning.* Washington, DC: National Academies Press. www.nap.edu/catalog.php?record_id=9596.

National Research Council (NRC). 2001a. *Educating teachers of science, mathematics, and technology: New practices for the new millennium.* Washington, DC: National Academies Press. www.nap.edu/catalog.php?record_id=9832.

National Research Council (NRC). 2001b. *Classroom assessment and the national science education standards.* Washington, DC: National Academies Press. www.nap.edu/catalog.php?record_id=9847.

National Research Council (NRC). 2001c. *Knowing what students know: The science and design of educational assessment.* Washington, DC: National Academies Press. www.nap.edu/catalog.php?record_id=10019.

National Research Council (NRC). 2006. *Systems for state science assessment.* Washington, DC: National Academies Press. www.nap.edu/catalog.php?record_id=11312.

National Research Council (NRC). 2007. *Taking science to school: Learning and teaching science in grades K–8,* ed. R. A. Duschl, H. A. Schweingruber, and A. W. Shouse. Washington, DC: National Academies Press. www.nap.edu/catalog.php?record_id=11625

National Research Council (NRC). 2010. *A framework for science education: Preliminary public draft.* Washington, DC: National Academies Press. The final version of this report will be available in 2011 from the Board on Science Education, Division of Behavioral and Social Sciences and Education: *http://www7.nationalacademies.org/bose.*

National Science Board (NSB). 2006. *America's pressing challenge—Building a stronger foundation.* Washington, DC: National Science Foundation. NSB 06-02. *www.nsf.gov/statistics/nsb0602.*

National Science Teachers Association (NSTA). 2000. The nature of science. NSTA Position Paper adopted in July 2000. Arlington, VA: NSTA. *www.nsta.org/about/positions/natureofscience.aspx.*

National Science Teachers Association (NSTA). 2003/Revised 2010. Standards for science teacher preparation. Arlington, VA: NSTA. *www.nsta.org/preservice.*

National Science Teachers Association (NSTA). 2004a. Science teacher preparation. NSTA Position Paper adopted in July 2004. Arlington, VA: NSTA. *www.nsta.org/about/positions/preparation.aspx.*

National Science Teachers Association (NSTA). 2004b. Science inquiry. NSTA Position Paper adopted in October 2004. Arlington, VA: NSTA. *www.nsta.org/about/positions/inquiry.aspx.*

National Science Teachers Association (NSTA). 2006. Professional development in science education. NSTA Position Paper adopted in May 2006. Arlington, VA: NSTA. *www.nsta.org/about/positions/profdev.aspx.*

National Science Teachers Association (NSTA). 2007a. Principles of professionalism for science educators. NSTA Position Paper adopted in June 2007. Arlington, VA: NSTA. *www.nsta.org/about/positions/professionalism.aspx.*

National Science Teachers Association (NSTA). 2007b. Induction programs for the support and development of beginning teachers of science. NSTA Position Paper adopted in April 2007. Arlington, VA: NSTA. *www.nsta.org/about/positions/induction.aspx.*

National Science Teachers Association (NSTA). 2007c. The integral role of laboratory investigations in science instruction. NSTA Position Paper adopted in February 2007. Arlington, VA: NSTA. *www.nsta.org/about/positions/laboratory.aspx.*

National Science Teachers Association (NSTA). 2008. The role of E-learning in science education. NSTA Position Paper adopted in September 2008. Arlington, VA: NSTA. *www.nsta.org/about/positions/e-learning.aspx.*

National Science Teachers Association (NSTA). 2010a. The role of research on science teaching and learning. NSTA Position Paper adopted in September 2010. Arlington, VA: NSTA. *www.nsta.org/about/positions/research.aspx.*

National Science Teachers Association (NSTA). 2010b. Teaching science and technology in the context of societal and personal issues. Arlington, VA: NSTA. *www.nsta.org/about/positions/societalpersonalissues.aspx.*

O'Brien, T. 1991. The science and art of demonstrations. *Journal of Chemical Education* 68 (11): 933–936.

O'Brien, T. 1992. Science inservice workshops that work for elementary teachers. *School Science and Mathematics* 92 (8): 422–426.

Research Cited

O'Brien, T. 1993. Teaching fundamental aspects of science toys. *School Science and Mathematics* 93 (4): 203–207.

O'Brien, T. 2000. A toilet paper timeline of evolution: A 5E cycle on the concept of scale. *American Biology Teacher* 62 (8): 578–582.

O'Brien, T. 2010. *Brain-powered science: Teaching and learning with discrepant events.* Arlington, VA: NSTA Press.

O'Brien, T., and D. Seager. 2000. 5 E(z) steps to teaching Earth-Moon scaling: An interdisciplinary mathematics/science/technology mini-unit. *School Science and Mathematics* 100 (7): 390–395.

Olenick, R. P. 2008. Comprehensive conceptual curriculum for physics (C3P) project: Misconceptions and preconceptions in introductory physics. *http://phys.udallas.edu/C3P/Preconceptions.pdf.*

Operation Physics. 2010. Children's misconceptions about science. *www.amasci.com/miscon/opphys.html.*

Osborne, R., and P. Freyberg. 1985. *Learning in science: The implications of children's science.* London: Heinemann.

Palmer, P. J. 2007. *The courage to teach: Exploring the inner landscape of a teacher's life.* 10th ed. San Francisco, CA: Jossey-Bass.

Posner, G. J., K. A. Strike, P. W. Hewson, and W. A. Gertzog. 1982. Accommodation of a scientific conception: Toward a theory of conceptual change. *Science Education* 66 (2): 211–227.

Sarquis, M., J. P. Williams, and J. L. Sarquis. 1995. *Teaching chemistry with toys: Activities for grades K–9.* New York: McGraw-Hill/Terrific Science Press.

Sarquis, M., L. Hogue, S. Hershberger, J. Sarquis, and J. Williams. 2009. *Chemistry with charisma: 24 lessons that capture and keep attention in the classroom.* Vol. 1. Middletown, OH: Terrific Science Press. *www.terrrificscience.org/charisma.*

Sarquis, M., L. Hogue, S. Hershberger, J. Sarquis, and J. Williams. 2010. *Chemistry with charisma: 28 more lessons that capture and keep attention in the classroom.* Vol. 2. Middletown, OH: Terrific Science Press. *www.terrrificscience.org/charisma.*

Schön, D. A. 1983. *The reflective practitioner: How professionals think in action.* New York: Basic Books.

Science Hobbyist, Amateur Science. Science myths in K–6 textbooks and popular culture. *www.amasci.com/miscon/miscon.html.* (This site includes extensive links to other sites.)

Shulman, L. 1986. Those who understand: Knowledge growth in teaching. *Educational Researcher* 15 (2): 4–14.

Shulman, L. 1987. Knowledge and teaching: Foundations of the new reform. *Harvard Educational Review* 57 (1): 1–22.

Singer, S. R., M. L. Hilton and H. A. Schweingruber. 2006. *America's lab report: Investigations in high school science.* Washington, DC: National Academies Press. (Executive Summary at *www.nap.edu/catalog/11311.html.*)

Stannard, C., T. O'Brien, and A. Telesca. 1994. STEP UP to networks for science teacher professional development. *Journal of Science Teacher Education* 5 (1): 30–35.

STEM Education Coalition. See especially links to Reports. *www.stemedcoalition.org.*

Tauber, R. T., and C. Sargent Mester. 2007. *Acting lessons for teachers: Using performance skills in the classroom.* 2nd ed. Westport, CT: Praeger.

Taylor, B., J. Poth, and D. Portman. 1995. *Teaching physics with toys: Activities for grades K–9.* New York: McGraw-Hill/Terrific Science Press.

Tobias, S., and A. Baffert. 2010. *Science teaching as a profession: Why it isn't. How it could be.* Arlington, VA: NSTA Press.

Treagust, D. F., R. Duit, and B. J. Fraser, eds. 1996. *Improving teaching and learning in science and mathematics.* New York: Teachers College Press.

Volkman, M. J., and S. K. Abell. 2003. Rethinking laboratories. *The Science Teacher* 70 (6): 38–41.

White, R. T., and R. F. Gunstone. 1992. *Probing understanding.* London: Falmer.

Wolpert, L. 1992. *The unnatural nature of science.* Cambridge, MA: Harvard University Press.

Yager, R. E. 1983. The importance of terminology in teaching K–12 science. *Journal of Research in Science Teaching* 20 (6): 577.

Yager, R. E., ed. 2005. *Exemplary science: Best practices in professional development.* Arlington, VA: NSTA Press.

Youngson, R. 1998. *Scientific blunders: A brief history of how wrong scientists can sometimes be.* New York: Carroll & Graf Publishers.

Index

A

A Framework for Science Education, Preliminary Public Draft, xix, 48, 84, 95, 120, 132, 144, 174, 181, 202–203
"A Terrible Test That Teaches" activity, xxviii, 263–278
   answers to questions in, 276–278
   debriefing for, 268–270
      with students, 269–270
      with teachers, 268
   expected outcome of, 263
   extensions of, 270–272
   Internet connections for, 272–276
   materials for, 265
   points to ponder for, 265
   procedure for, 266–267
   science concepts in, xxviii, 264
   science education concepts in, 264
AAAS (American Association for the Advancement of Science), xviii, xix, 9, 87, 91, 144, 154, 168, 202
Accessibility, as source of learner misconceptions, 300
Acid-base indicators, 143–152
Acronyms, xxv
AIDS prevention, 224, 227, 234
Allergies, xxx, 5
American Association for the Advancement of Science (AAAS), xviii, xix, 9, 87, 91, 144, 154, 168, 202
American Chemical Society, 149
American Society for Microbiology, 214
Ammonia, 143–147, 149
An Inconvenient Truth, 254
Anagrams, 111, 112, 116
Analogy-based simulation models, 24, 238
Answers to embedded questions, xxii, xxiii
Archimedes, 157, 162–165, 167–168, 171
"Archimedes, the Syracuse (Sicily) Scientist" activity, xxvii, 161–172
   answers to questions in, 170–172
   debriefing for, 165–166
      with students, 166
      with teachers, 165–166
   expected outcome of, 161
   extensions of, 166–168
   Internet connections for, 168–169
   materials for, 162–163
   points to ponder for, 163
   procedure for, 164–165
   safety note for, 167
   science concepts in, xxvii, 162
   science education concepts in, 162
Art and science of science education, xi–xii
Art of Teaching, The, xi
Asimov, Issac, 61
Assessments, xv, xx–xxi, xxviii, 92, 263–278
   alternative, xxi, 270
   diagnostic, 264, 277, 279–294, 304
   formative, 264, 277–278, 280, 288, 289, 297, 304
   international, 122–123
   misconceptions about, xv, xx–xxi
   summative, 264, 277, 278, 280, 288, 289, 297
Assimilation-type learning, 303
Association of Biology Teachers, 101
Asthma, xxx
Attitudes about science, 289
Ausubel, David, xii
Avogadro's number, 212

B

Bacon, Roger, 155
Barrow, John, 210
BBS (black box system), xxv, 4, 48, 50, 72, 75–77, 81, 84
ben Maimon, Moses, 60
Benchmarks for Science Literacy, xviii, xix, 9, 48, 84, 91, 95, 120, 132, 137, 144, 154, 162, 165, 168, 174, 181, 202–203, 209
Bernal, John D., 163
Best-practice teaching, xiv, 95
Big ideas in science, xiv, xix, 18, 23, 94, 99, 147
Biological analogies and applications (BIO), xxv
Black box system (BBS), xxv, 4, 48, 50, 72, 75–77, 81, 84
Blake, William, 96, 98

Index

Bloom, Benjamin, 265, 268
Bloom's taxonomy, 268, 270, 278, 280, 297
Bosch, Carl, 149–150
Botany, 285–287, 292–294
"Bottle Band Basics" activity, xxvii, 189–199
   answers to questions in, 198–199
   debriefing for, 193
      with students, 193
      with teachers, 193
   expected outcome of, 189
   extensions of, 194–196
   Internet connections for, 196–198
   materials for, 190
   points to ponder for, 191
   procedure for, 191–192
   science concepts in, xxvii, 190
   science education concepts in, 190
Bruner, Jerome, xii
Buoyancy
   density and, 33–46
   water displacement and, 162, 165, 166–167, 171–172

C

Campbell, Thomas, 145
Carbon footprint, 254, 260
Carroll, Lewis, 20, 86, 121, 122, 271
CFLs (compact fluorescent lightbulbs), 247–260
Chemical & Engineering News, 149
Chemical warfare, 150
Chemicals, safe use of, xxix–xxx
CIA (curriculum-instruction-assessment) practices,
   xiii–xiv, xv, xxi, 6, 10, 38–39, 84, 122, 123, 132, 137,
   144, 154, 162, 190, 203, 248, 281, 302–303
Classroom safety practices, xxix–xxx
Cognitive disequilibrium, xv, 108
Cognitive learning theory and research, xii, xv, 58
Color-changing inks, 148–149
"Comeback Cans" activity, xvii, xxvi, 3–16
   answers to questions in, 14–16
   debriefing for, 9–10
      with students, 10
      with teachers, 9–10
   expected outcome of, 3
   extensions of, 11–12
   Internet connections for, 12–14
   materials for, 4–5
   points to ponder for, 6
   procedure for, 6–8
      with students, 7–8
      with teachers, 6–7
   safety notes for, 4, 5
   science concepts in, xxvi, 4
   science education concepts in, 4
Common Core State Standards initiative, 120
Compact fluorescent lightbulbs (CFLs), 247–260
Compasses, 119–124
Concept mapping, 78, 203, 304
Conceptual accommodation, 303
Constructivist learning theory, xii, 59, 78, 84
"Cookie Mining" activity, xxviii, 237–246
   answers to questions in, 245–246
   debriefing for, 241–242
      with students, 241–242
      with teachers, 241
   expected outcome of, 237
   Internet connections for, 243–244
   materials for, 238
   points to ponder for, 239
   procedure for, 239–240
   safety notes for, 238, 239
   science concepts in, xxviii, 238
   science education concepts in, 238
Courage to Teach, The, xi, 176, 180
Crossword puzzles, 77
Curriculum standards, xix, xxvii, 48, 84, 95, 120, 132,
   136–137, 144, 154, 162, 165, 168, 174, 181, 202–203
Curriculum-instruction-assessment (CIA) practices,
   xiii–xiv, xv, xxi, 6, 10, 38–39, 84, 122, 123, 132, 137,
   144, 162, 190, 203, 248, 281, 302–303

D

da Vinci, Leonardo, 155
Debriefing, xxii
Democritus, 96
Demonstration-experiments, xii, xiv, 303
Density and buoyancy, 33–46
Dewey, John, xii
"Diagnostic Assessment" activity, xxviii, 279–294
   answers to questions in, 291–294
   debriefing for, 287–289
      with students, 289
      with teachers, 287–288
   expected outcome of, 279
   extensions of, 289–290
   Internet connections for, 290
   materials for, 281
   points to ponder for, 282
   procedure for, 283–286

science concepts in, xxviii, 280
science education concepts in, 280–281
Discrepant-event activities, xi, xii, xiii, xiv–xv, 303
   acronyms used in, xxv
   anomalous outcomes of, xv
   in biological sciences, 309
   to develop process and reasoning skills, 311–312
   in Earth and physical sciences, 310–311
   format of, xxii–xxiv
   probing questions for, xxiii
   purposefully puzzling, xv
   purposes of, xi, xiv–xv, xxii
   science concepts covered in, xv, xxvi–xxviii
   science content and process skills covered in, 309–312
   S_2EE_2R criteria for, xvii, xxii
   time required for, xxiii–xxiv
Discussions, data-based, xiv
Disease transmission, 223–235
"Dual-Density Discrepancies" activity, xvii, xxvi, 33–46
   answers to questions in, 45–46
   debriefing for, 38–40
      with students, 39–40
      with teachers, 38–39
   expected outcome of, 33
   extensions of, 40–43
   Internet connections for, 43–45
   materials for, 35
   points to ponder for, 36
   procedure for, 36–38
   safety notes for, 35, 41, 43
   science concepts in, xxvi, 33
   science education concepts in, 34–35
DuBos, Rene J., 176, 225, 227

E

Eames, Charles, 205
Educational philosophers, xii
Educational psychologists, xii
Einstein, Albert, 6, 60, 73, 109, 124, 265, 268, 282, 288
Electromagnetism, 124
Eliot, T. S., 191
Empirical evidence, 18, 21, 22, 50, 58, 62–63, 74, 80, 84–87, 94, 104–105, 108, 122, 162, 165, 172
Environmental conservation, 237–246, 252, 254, 259–260
Environmental Protection Agency, 259
European Union, 260
Everyday experiences, as source of learner misconceptions, 299

Expected outcome, xxii
Extensions, xxii, xxiii
Eye protection, xxix

F

Faraday, Michael, xi, 49, 124
Feynman, Richard, xii, 49, 61, 176
5E Teaching Cycle, xxiii, 10, 75, 76, 99, 148, 280, 288, 289, 304
"Follow That Star" activity, xxvii, 119–129
   answers to questions in, 127–129
   debriefing for, 122–124
      with students, 123–124
      with teachers, 122–123
   expected outcome of, 119
   extensions of, 124–125
   Internet connections for, 125–127
   materials for, 121
   points to ponder for, 121
   procedure for, 122
   science concepts in, xxvii, 120
   science education concepts in, 120
Ford, Henry, 249
Formative assessments, 264, 277–278, 280, 288, 289, 297, 304
Franklin, Ben, 194
Future research, as source of learner misconceptions, 301–302

G

Galilei, Galileo, 155, 157, 168, 265, 268
Geisel, Theodore, 239
Geocaching, 125
Gilbert, William, 124
Global warming, 250, 252, 254
Gore, Al, 254
Gould, Stephen Jay, 61, xii, 109
GPS devices, 120–129
Grading, xx–xxi, 264, 277, 283, 289. *See also* Assessments

H

Haber, Fritz, 149–150
Hakin, Joy, 287
Hands-on exploration (HOE), xiv, xv, xxv
Harry Potter books, 147
Hawking, Stephen, 61

Index

Health and hygiene, 223–235
Henry, Joseph, 124
Higgins, Sir William, 145
History of science (HOS), xxv, 162, 167–168
HIV transmission, 224, 227, 234
HOE (hands-on exploration), xiv, xv, xxv
"Horsing Around" activity, xxvii, 131–141
    answers to questions in, 141
    debriefing for, 136–137
        with students, 137
        with teachers, 136–137
    expected outcome of, 131
    extensions of, 137–140
        for students, 138–140
        for teachers, 137–138
    Internet connections for, 140–141
    materials for, 132
    points to ponder for, 133
    procedure for, 134–136
    science concepts in, xxvii, 132
    science education concepts in, 132
HOS (history of science), xxv, 162, 167–168
Houbert, Joseph, 36
Hubble, Edwin Powell, 191
Huxley, Thomas Henry, xi, 49, 60, 121
Hydrometer, 43

I

Infectious disease transmission, 223–235
"Inferences, Inquiry, and Insight" activity, xxvi, 47–55
    answers to questions in, 55
    debriefing for, 51–52
        with students, 52
        with teachers, 51–52
    expected outcome of, 47
    extensions of, 52–53
    Internet connections for, 54–55
    materials for, 49
    points to ponder for, 49
    procedure for, 50–51
    science concepts in, xxvi, 48
    science education concepts in, 48–49
Inquiry-oriented science lessons, xii, xiv, xvi, xxiii–xxiv
Instructional scaffolding, 304
International science assessments, 122–123
International System of Units, 202, 206
Internet connections, xxii, xxiii

J

James, William, 191
Jefferson, Thomas, 49
Jigsaw puzzles, 113

K

Kepler, Johannes, 59
Kinetic molecular theory (KMT), 94, 99, 174

L

Lack of prior experiences, as source of learner misconceptions, 299
Lanier, Judith, 271
Latex allergy, xxx
Learner-Centered Curriculum and Assessment for New York State, 133, 282
Learning
    assimilation-type, 303
    cognitive theory of, xii, xv
    by conceptual accommodation, 303
    lifelong, xiii, xiv
    providing adequate time for, 94–95
    teachers' role in, xiii–xiv
Learning communities, xiii, xiv
Learning progressions, 99, 304
    misconceptions due to concept placement in, 300
Learning-teaching dynamic, 295
Levers, 162–166, 170–171
Lightbulbs, 247–260
Linear units, 206, 218
Locke, John, 96
Logical arguments, 18, 21, 22, 50, 58, 63, 74, 80, 84–87, 94, 97, 105, 108, 122, 155, 162, 165, 172
Lord Kelvin, 265, 268

M

Magazines, pseudoscience articles in, 57–70
"Magic Bus of School Science" activity, xxvii, 83–92
    answers to questions in, 91–92
    debriefing for, 87–88
        with students, 87–88
        with teachers, 87
    expected outcome of, 83
    extensions of, 88–89
    Internet connections for, 89–91
    materials for, 85
    points to ponder for, 86

procedure for, 85–87
science concepts in, xxvii, 84
science education concepts in, 84–85
"Magical Signs of Science" activity, xxvii, 143–152
    answers to questions in, 152
    debriefing for, 147–148
        with students, 148
        with teachers, 147–148
    expected outcome of, 143
    extensions of, 148–150
    Internet connections for, 150–151
    materials for, 144–146
    points to ponder for, 145
    procedure for, 146–147
    safety notes for, 145, 146
    science concepts in, xxvii, 144
    science education concepts in, 144
Magnets, 124
"Making Sense by Spending Dollars" activity, xxviii, 247–260
    answers to questions in, 258–260
    debriefing for, 252–253
        with students, 253
        with teachers, 252–253
    expected outcome of, 247
    extensions of, 253–255
    Internet connections for, 255–258
    materials for, 249–250
    points to ponder in, 249
    procedure for, 250–252
        with students and teachers, 250–252
        with teachers only, 250
    science concepts in, xxviii, 248
    science education concepts in, 248
Material Safety Data Sheets (MSDS), xxix
Materials, xxii
Mathematics
    metric measurements, 157–160, 201–220
    of music, 195
    volume measurements, 153–160
"Measurements and Molecules Matter" activity, xxvii, 173–187
    answers to questions in, 185–187
    debriefing for, 179–181
        with students, 181
        with teachers, 179–181
    expected outcome of, 173
    extensions of, 181–183
    Internet connections for, 184–185
    materials for, 174–175
    points to ponder for, 176
    procedure for, 177–179
    safety notes for, 175, 187
    science concepts in, xxvii, 174
    science education concepts in, 174
"Medical Metaphor Mixer" activity, xxviii, 223–235
    answers to questions in, 234–235
    debriefing for, 227–228
        with students, 227–228
        with teachers, 227
    expected outcome of, 223
    extensions of, 228–231
    Internet connections for, 232–234
    materials for, 225
    points to ponder for, 225
    procedure for, 226–227
        with teachers, 226–227
    safety note for, 225
    science concepts in, xxviii, 224
    science education concepts in, 224
"Metric Measurements, Magnitudes, and Mathematics" activity, xxvii, 201–220
    answers to questions in, 218–220
    debriefing for, 209–210
        with students, 210
        with teachers, 209–210
    expected outcome of, 201
    extensions of, 210–215
    Internet connections for, 216–218
    materials for, 203–204
    points to ponder for, 205
    procedure for, 205–208
    science concepts in, xxvii, 202
    science education concepts in, 202–203
Microbes and human health, 223–235
Mining and resource management, 237–246
Misconceptions of learners, xiii, xv, xxi–xxii, xxiii, 10, 39, 144, 277, 280, 281, 295–304
    concept-building approaches to, 303–304
    perception of discrepant events, 302–303
    sources of, 297–302
        accessibility, 300
        everyday experiences, 299
        future research, 301–302
        lack of prior experiences, 299
        misunderstandings about science, 300–301
        placement in learning progression, 300
        popular culture, 297–298
        religious conflicts, 301
        school culture and artifacts, 298

Index

translation errors, 298–299
Misunderstandings about science, 300–301
Mixer (MIX), xxv
MSDS (Material Safety Data Sheets), xxix
Muir, John, 205
Music and sound, 189–199

N

Nanoscale science, 215
Nation at Risk, 120
National Research Council (NRC), xviii, 9, 91, 144, 168
National Science Education Standards (NSES), xi, xix, 9, 48, 84, 86–87, 91, 95, 120, 132, 136, 144, 162, 165, 168, 174, 180, 202–203, 209, 249
National Science Teachers Association (NSTA), xi, xiv, xviii, xix, 101
    Standards for Science Teacher Preparation, xvi, xix
Nature of science (NOS), xviii, xxv, xxvi–xxvii, 4, 18, 48, 58, 120, 132, 144, 281
NCLB (No Child Left Behind) Act, 120, 268, 271
NDT (nondestructive testing), 9
Neuroplasticity, 295
Newspapers
    biomolecular polymers composing, 93–105
    comic strips in, 100
    pseudoscience articles in, 57–70
    recycling of, 100–102
Newton, Isaac, 60, 73, 168, 176
Nitrates, 243
No Child Left Behind (NCLB) Act, 120, 268, 271
Nobel, Alfred, 150
Nondestructive testing (NDT), 9
NOS (nature of science), xviii, xxv, xxvi–xxvii, 4, 18, 48, 58, 120, 132, 144, 281
NRC (National Research Council), xviii, 9, 91, 144, 168
NSES (National Science Education Standards), xi, xix, 9, 48, 84, 86–87, 91, 95, 120, 132, 136, 144, 162, 165, 168, 174, 180, 202–203, 209, 249
NSTA (National Science Teachers Association), xi, xiv, xviii, xix, 101
    Standards for Science Teacher Preparation, xvi, xix

O

Oersted, Hans Christian, 124

P

PAD (participant-assisted demonstration), xxv
Palmer, Parker J., 176
Paper-and-pencil puzzles (PPPs), xviii, xxv, 48–55, 84
Participant-assisted demonstration (PAD), xxv
Participatory analogy, 264
Pasteur, Louis, 249
Pattern finding, 109–112, 115–116
Paulos, John Allen, 209
Pavlov, Ivan, 86, 87, 91
PCK (pedagogical content knowledge), xi, xiii, xvi, 272
Peanut allergy, xxx, 5
Pedagogical content knowledge (PCK), xi, xiii, xvi, 272
Phenolphthalein "ink," 143–145
Pi, 157–158, 160
Piaget, Jean, xii
PISA (Program for International Student Assessment), 122–123
Pitch, 190, 192–195, 198–199
Plants, 285–287, 292–294
POE (predict-observe-explain) activities, xxv, 4, 194, 195, 277, 303
Points to ponder, xxii
"Pondering Puzzling Patterns and a Parable Poem" activity, xxvii, 107–116
    answers to questions in, 115–116
    debriefing for, 111–112
    expected outcome of, 107
    extensions of, 113–114
    Internet connections for, 114–115
    materials for, 109
    points to ponder for, 109
    procedure for, 109–111
    science concepts in, xxvii, 108
    science education concepts in, 108–109
Popper, Karl, 60
Popular culture, as source of learner misconceptions, 297–298
PPPs (paper-and-pencil puzzles), xviii, xxv, 48–55, 84
Preconceptions. *See* Misconceptions of learners
Predict-observe-explain (POE) activities, xxv, 4, 194, 195, 277, 303
Priestly, Joseph, 149
Procedure, xxii
Process and reasoning skills for teaching-learning science, 311–312
Professional development, xi–xvi, xxiii–xxiv
Program for International Student Assessment (PISA), 122–123
Protective clothing, xxix
"Pseudoscience in the News" activity, xxvi, 57–70
    answers to questions in, 68–70

        debriefing for, 61–62
            with students, 61–62
            with teachers, 61
        expected outcome of, 57
        extensions of, 62–65
        Internet connections for, 65–67
        materials for, 59
        points to ponder for, 60
        procedure for, 59–61
        science concepts in, xxvi, 58
        science education concepts in, 58–59
Psychological rewards, 108
Puzzles, xxvii, 107–116, 131–141
        crossword, 77
        jigsaw, 113
        paper-and-pencil, xviii, xxv, 48–55, 84
Pythagoras of Samos, 195
Pythagorean theorem, 159, 210

R

"Reading Between the Lines of the Daily Newspaper"
        activity, xxvii, 93–105
        answers to questions in, 104–105
        debriefing for, 98–100
            with students, 99–100
            with teachers, 98–99
        expected outcome of, 93
        extensions of, 100–102
        Internet connections for, 102–104
        materials for, 95
        points to ponder for, 96
        procedure for, 96–98
        safety note for, 95
        science concepts in, xxvii, 94
        science education concepts in, 94–95
Real-world science instruction, xix–xx, xxviii, 108
Recycling, 100–102, 242, 260
Rees, Sir Martin, 86
Religious conflicts, as source of learner
        misconceptions, 301
Resonance, 194
Resource management, 237–246
Reverse pattern finding, 52

S

Safety practices, xxix–xxx
Sagan, Carl, xii, 21, 60, 61, 287
Saxe, John Godfrey, 113

Scales of science, 210–213
Scalia, Supreme Court Justice Antonin, 250
School culture and artifacts, as source of learner
        misconceptions, 298
Science concepts, xv, xviii, xxii, xxvi–xxviii, 180
Science content and process skills, 309–312
Science education analogies, xi, xii, xv
Science education concepts, xviii, xxii
Science education topics, xxv–xxviii
Science writing, 57–70
Science-technology-engineering-mathematics (STEM)
        education, xix–xx, 58, 248, 250, 252
Science-technology-society (STS) issues, xix–xx, xxv,
        xxviii, 42–43, 84, 88, 94, 100–102, 124–125, 149–
        150, 182–183, 196, 224, 227–228, 230–231, 238,
        248, 250, 252, 254–255
Scientific literacy, 58, 248–250
"Scientific Reasoning" activity, xxvi, 71–81
        answers to questions in, 80–81
        debriefing for, 74–75
            with teachers, 75
            with teachers and students, 74–75
        expected outcome of, 71
        extensions of, 75–78
        Internet connections for, 78–79
        materials for, 73
        points to ponder for, 73
        procedure for, 74
        science concepts in, xxvi, 72
        science education concepts in, 72–73
Scientific reasoning assessment, 284, 291–292
Scientific validity, triple test of, 21, 22, 62–64, 94, 123
Scientist-teachers, xi–xii
S_2EE_2R criteria, xvii, xxii
S_2EE_2R demonstration analysis form, xxii, 305–308
Seesaw balance activity, 164, 170–171
Seneca, Lucius Annaeus, 36
SI units, 202, 206
Simulation models, 224, 238
Skeptical review, 18, 21, 22, 50, 58, 59, 63–64, 74, 80,
        84–87, 94, 98, 105, 108, 123, 156, 162, 165, 172
Socrates, xii
Sound, 189–199
Spencer, Herbert, 121
STEM (science-technology-engineering-mathematics)
        education, xix–xx, 58, 248, 250, 252
Stevenson, Adlai, 239
STS (science-technology-society) issues, xix–xx, xxv,
        xxviii, 42–43, 84, 88, 94, 100–102, 124–125, 149–
        150, 182–183, 196, 224, 227–228, 230–231, 238,

Index

248, 250, 252, 254–255
Students
   allergies of, xxx, 5
   attitudes about science, 289
   perception of discrepant events by, 302–303
   science misconceptions of, xiii, xv, xxi–xxii, xxiii, 10, 39, 144, 277, 280, 281, 287, 292, 295–304
Summative assessments, 264, 277, 278, 280, 288, 289, 297
Surface area units, 206–207, 218–219
Surface area-to-volume ratios, 213–214
Szent-Gyorgyi, Albert, 109, 114

T

Teacher demonstration (TD), xxv
Teachers
   assessing performance of, 272
   as lifelong learners, xiv, xxiv
   professional development of, xi–xvi, xxiii–xxiv
   professional learning communities of, xiv
   role in student learning, xiii–xiv
   science misconceptions of, 280, 297
Television science programs, 88, 287
Terrific observations and yearnings for science (TOYS), xxv, 9–11
"The Unnatural Nature and Uncommon Sense of Science" activity, xvii, xxvi, 17–29
   answers to questions in, 27–29
   debriefing for, 22–23
      with students, 22–23
      with teachers, 22
   expected outcome of, 17
   extensions of, 23–24
   Internet connections for, 25–27
   materials for, 19
   points to ponder for, 20–21
   procedure for, 19–22
   science concepts in, xxvi, 18
   science education concepts in, 18–19
Thomas, Lewis, 61
Thompson, Francis, 96, 98
Thompson, William, 265, 268
TIMSS (Trends in International Mathematics and Science Study), 122
TOYS (terrific observations and yearnings for science), xxv, 9–11
Toys in Space program, 16

Translation errors, as source of learner misconceptions, 298–299
Trends in International Mathematics and Science Study (TIMSS), 122
Triple test of scientific validity, 21, 22, 62–64, 94, 123
Tuning fork, 194
Tyndall, John, 163

V

Ventilation system, xxx
"Verifying Vexing Volumes" activity, xxvii, 153–160
   answers to questions in, 160
   debriefing for, 156–157
      with students, 157
   with teachers, 156
   expected outcomes of, 153
   extensions of, 157–159
   Internet connections for, 159
   materials for, 154
   points to ponder for, 155
   procedure for, 155–156
   safety note for, 154
   science concepts in, xxvii, 154
   science education concepts in, 154
Verne, Jules, 239
Visual participatory analogies, xv, xvii, xix, xxiii
Volume measurement, 153–160, 201–220
Volume reduction when mixing liquids, 173–187
von Goethe, Johann Wolfgang, 121
Vygotsky, Lev, xii

W

Waksman, Selman Abraham, 225
Ward, William Arthur, 36
Water displacement and buoyancy, 162, 165, 166–167, 171–172
Whitehead, Alfred North, 6, 10, 133, 137
Wilson, E. O., 61
Wolpert, Lewis, 20
Wordsworth, William, 145

Z

Zone of proximal development (ZPD), 280